1950s Plastics Design

# Everyday Elegance

with Price Guide

"Flight to the Future"

...TO THE WORLD OF PLASTICS

Holly Wahlberg

Schiffer Publishing Ltd

Copyright © 1994 by
Holly Wahlberg

Library of Congress Catalog Number: 93-87053
All rights reserved. No part of this work may be repro-
duced or used in any forms or by any means – graphic,
electronic or mechanical, including photocopying or
information storage and retrieval systems – without
written permission from the copyright holder.

Printed in the United States of America.
ISBN: 0-88740-563-0

Published by Schiffer Publishing, Ltd.
77 Lower Valley Road
Atglen, PA 19310
Please write for a free catalog.
This book may be purchased from the publisher.
Please include $2.95 postage.
Try your bookstore first.

We are interested in hearing from authors
with book ideas on related subjects.

# Contents

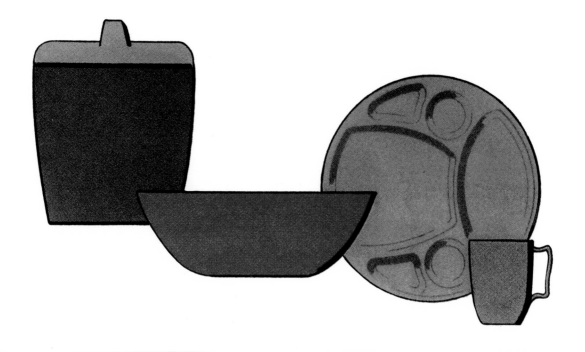

Wonder, awe even kneeling before the "miracle" of vinyl plastic.

All things seemed possible for newlyweds of the 1950s whose wedded beginnings paralleled the expansion of modern technology - even to the moon.

Higher levels of disposable income and an innocent trust in advertisers' claims sent consumers on a plastics buying spree.

# Chapter 1:
## Better Living ...Through Chemistry

Throughout the 1950s, headlines announced that the whole world was "going plastic." A "revolution" in consumer products was occurring thanks to advances in the new "wonder material," plastic. In this rhetorical climate, consumers could be excused for believing they were witnessing a modern miracle, and in a sense, they were. The technological advances brought about by the exigencies of World War II created a new generation of postwar plastics that did indeed seem "revolutionary" - capable of bringing "carefree" beauty into new areas of the home beyond just the kitchen and bathroom.

With their mixture of glamour and practicality, plastics met the needs and aspirations of the informal, child-centered postwar American culture. Within a single generation, plastics rose to astonishing sales and production levels. And although this rise was not without serious setbacks, the success of plastics during the 1950s was inarguably staggering.

Relaxed entertaining represented an effort to achieve a new spontaneity in American life.

Advances in plastics linked the "magic of chemistry" to the needs of postwar America.

The triumph of plastic was closely tied to "gracious living," that 1950s phrase for a generation's belief that happiness was to be found in a more relaxed lifestyle centered around growing families, outdoor living and informal entertaining. The affordable mass production of new miracle materials such as plastic seemed to make "gracious living" possible even for ordinary people. The belief that all would now have a chance to own stylish, functional and affordable products was a source of considerable national pride in America, and the concept of gracious living quickly took on a larger symbolic importance in 1950s rhetoric. Many believed that the American honesty, ingenuity and generosity embodied in "gracious living" would lead the rest of the world in postwar

Plastics combined practicality and economy with the appeal of lively colors.

New postwar plastics were especially appropriate for families starting a new way of life in fledgling suburban developments.

life - not just commercially and politically, but domestically as well. In 1954, for example, *House Beautiful* boasted that, "We are achieving comfort, beauty, personal meaning on a scale that no other civilization has ever enjoyed - And we are achieving it where it most counts: in the smallest details of our daily lives. The quality of daily life is the finest test we know of the quality of any civilization."

Americans could take pride in this new higher quality of daily life, in large part, because of advances in the field of consumer plastics. Colorful, affordable plastics seemed to promise a way for Americans to have gaiety, sparkle and individuality even within the uniformity of suburban tract housing. And in an era when servants would vanish from daily life, the easy maintenance of "wipe clean" plastic surfaces promised to eliminate the time consuming and dull household chores which kept a housewife away from her husband and children.

Advertising which stressed plastic's low maintenance found a receptive audience among housewives unfulfilled by domestic work. However, the "quick swish of a cloth" housework plastic provided often only increased the emptiness and boredom of a housewife's day and intensified the pressure to use her increased "leisure" to mother "perfect" children.

6

Advertisers portrayed plastics as the great liberator of housewives everywhere. Many women wondered, however, what came after the "liberation."

Amazing Plastic Floor!

CUTS CLEANING CARE! up to 40%

Want a miracle floor—a sparkling plastic wonder that *glows* with beauty and laughs at dirt? That's Flor-Ever, the new Vinylite creation that stays bright with less, less waxing—and is cleaned far, far faster and *easier*. Why is Flor-Ever so wonderful? It's NON-POROUS...greasy grit and dulling film can't get a grip on this slick, super-smooth surface — and *nothing used in kitchens can stain Flor-Ever!* (Think, too, how grand Flor-Ever will be on kitchen sinks and counters.)

Naugahyde Protection*

As easy as washing a dinner plate!

This cozy scene tried to demonstrate the degree of family togetherness possible when living rooms were armored with plastic products.

"Don't worry, it's Boontonware!"

Guaranteed by Good Housekeeping

All the family can relax with Boontonware!

Plastic dishes were depicted as protectors of the mother-daughter relationship.

8

In the sphere of family life, plastic fit in perfectly with the decade's emphasis on creativity, fun and "togetherness" in child rearing. For the first time, children could be constantly present and constantly creative in rooms "armored" in plastic products. There would be no more "adult only" rooms comprised of furnishings too costly, luxurious and perishable to withstand rough play, spills and crayons. In many 1950s homes, a basement family room may have been set aside for particularly rough play or "teen dance parties." But for the most part, all areas of the house were open to children and their sometimes destructive play thanks to the strength of plastic materials. Technological advances over the course of the decade even made "good looks" possible for plastic. Claims of luxury and beauty began to soften plastic's no-nonsense utilitarian image.

Once plastics had proven their practicality, an air of luxury - even romance and beauty - was added to increase sales.

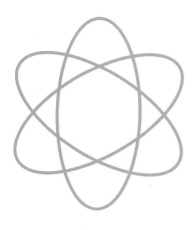

a floor

is moonlight...

and romance...

and beauty...

The backyard patio was a place for family relaxation and casual parties in a natural setting.

Plastic also proved essential in creating the California-style "outdoor living" so admired by a generation which aspired to spontaneity and healthfulness in its entertaining and leisure activities. Terraces and patios were furnished with sunfast plastic covered furniture, and guests ate from plastic dinnerware. The "happy crowd" enjoying the "wholesome out of doors" was a cherished symbol of the new era, and plastic was central in achieving this form of recreation.

The huge range of 1950s plastic products meant there was something for everyone in the plastics field, and growth of plastics consumption was phenomenal. Whether buying the sleekest embodiments of modern sculptural design or the most utilitarian do-it-yourself kitchen counters, plastic was often the material of choice. It could imitate traditional materials such as leather, brick, cloth, china, wood, even metal. Or it could embody boldly flowing organic forms achievable only at great expense and difficulty in other materials. Plastic's sleek, trim lines and mold perfect finish brought a sense of clean, fresh orderliness

Throughout the decade, plastic expanded into all rooms of the house in a wide range of applications.

that helped make tiny apartments or cramped ranch houses seem bigger. Its integral color eliminated the chipping common in painted wood surfaces, while its rigidity and strength meant a lack of the corrosion and denting so troublesome in metal products.

From a manufacturer's point of view, plastic was relatively inexpensive to produce. Each product was uniform without much waste or hand labor. Even scrap could be reused. And plastic proved highly adaptable to a whole variety of production techniques. The most complicated product housings were possible, and assembly costs were minimal. Such things as knob openings and joint fastenings could be molded right into the product rather than applied by hand afterwards.

These manufacturing advantages, coupled with a special applicability to the social needs of the 1950s, made plastic the ideal material for a generation that aspired to glamour and luxury combined with durable practicality. In their price, plastics promised to democratize luxury. In their toughness, they seemed to make glamour sensible.

Some plastic products, like this lazy susan and candy dish, had multiple pieces but were assembled without screws or soldering since plastic could be designed with joint mechanisms already molded into the finished product.

In the late 1940s, consumers faced a barrage of disappointing plastic products from ignorant, careless or unscrupulous manufacturers. In this series of American Cyanamid ads from 1947 and 1948, consumers encounter scorched buttons, flammable stove handles and melting cups.

*A cup of woe for Mrs. Doe...*

# Chapter 2:
# Origins of 1950s Plastics

Considering their well established position in today's world, the synthetic plastics developed during and just prior to World War II had a surprisingly rocky beginning in the consumer marketplace. Plastics were utilized during the first years of the war as substitutes for consumer goods made of traditional materials such as wood, metal, leather, and glass. Soon though, plastics too were made part of the military production effort, and plastics for consumer goods grew scarce. But as the war drew to a close, demand grew once again, particularly when plastics were presented to war-weary consumers as having miraculous properties symbolizing the new era about to dawn. This overselling of plastics to overeager consumers helped create the mistaken impression that "plastic" was a single miracle substance which could be used in any situation for any product.

Ignorance of plastic's complexity extended even to manufacturers. The plastics industry had grown so quickly that many manufacturers were little better off than consumers when it came to understanding the highly technical nomenclature of the industry. One business analyst of the period ominously warned in 1950, "The plastics business is clearly no longer one for novices. It's a highly technical field, and to succeed in it, you need a terrifying amount of special knowledge."

But the structure of the postwar plastics industry involved approximately fifty raw material producers and thousands of small scale companies that molded plastics into products. Many of these small companies had little knowledge about the properties of the particular plastic with which they were working. Unfortunately, plastics were understood by some molders as simple, interchangeable substitutes for whatever natural material was unavailable. Consequently, the wrong type of plastic was frequently chosen for a job, one it was incapable of doing. To make matters worse, the hapless consumer hadn't a clue about what to expect from or how to care for each of plastic's many varieties.

Articles appeared in business and home magazines trying to sort out this confusion. One writer patiently explained the problem this way: "Perhaps the consumer happens to kick a polystyrene toy under the radiator. The toy melts out of shape. Henceforth, the consumer harbors a deep distrust of all plastics. He doesn't know there are many other types that won't melt under a radiator. The metals industries, being much older, escape this kind of trouble. The fact that copper occasionally turns green has no influence on the consumer's regard for other metals. He knows that other metals behave in other ways. It's a matter of familiarity."

Ignorance was only one factor at work, however. By 1946, unscrupulous molding companies were springing up in the hope of making a quick profit in what was clearly a seller's market. Out of this came an array of sleazy shower curtains, melting buttons, thin walled cups, and wilting razors. By 1948, plastics sales had plummeted, and disgruntled consumers considered plastic more of a swindle than a miracle.

Realizing immediate action had to be taken, the Society for the Plastics Industry and raw material suppliers began to set quality standards that would effectively police the industry. Such corporate giants as duPont, Monsanto, American Cyanamid, Catalin, and Bakelite developed a paternalistic attitude toward their fabricator clients, helping them with marketing campaigns and making advisory panels of their leading experts available for technical consultations.

But encouraging better quality workmanship and a greater understanding of each type of plastic were only the beginnings of necessary reform. The industry was accustomed to defending plastic as much more than a cheap, shoddy, "substitute" material. But the low cost, high speed and relative ease of molding the new plastics made the temptation for quick profit and sloppy standards greater than it had ever been before. The permanent discrediting of plastic could only be prevented by educating both the consumer and the molder. Plastic's fate in the peacetime economy depended on it. Only through masterful marketing and a unique set of historical circumstances did plastic become the absolute right material for a new generation of Americans.

Melamine dinnerware in particular exemplifies the success plastics came to enjoy in the 1950s. The speed and extent of melamine's entrance into 1950s daily life was astonishing. Twenty-five percent of all American households, over ten million families, owned at least one set of melamine dishes by 1956, just seven years after the first dishes appeared on the domestic market.

Plastic as a material for dishes was not, however, a totally new concept. Throughout the late 1920s and 1930s, a urea formaldehyde plastic dinnerware was developed for domestic use by the American Cyanamid Corporation, which set out to create an American version of British picnic dishes known as Beetleware. Americans first came into contact with this urea formaldehyde plastic through radio program and cereal box giveaways advertising Wheaties, Ovaltine and Bisquick. In the early 1930s, five million Wheaties cereal "skippy bowls" and over four million Little Orphan Annie Ovaltine "shake-up mugs" introduced Americans to this still relatively unknown type of plastic.

But American Beetleware, although cheap to produce with injection molding, faded and cracked after repeated contact with water. Its fillers (the material added to the plastic to make it more workable) were usually wood flour or pulp which broke down over time when in contact with water. This trait made it an unsuitable material for dinnerware even though urea formaldehyde was light colored, inexpensive, attractive, and ideal for producing finely molded details. So although urea plastic dishes were accepted for an occasional outdoor breakfast or picnic, they never appeared at the dinner table.

Plastic premiums, like the 1932 Wheaties cereal "skippy bowl," were a small preview of the inexpensive plastic houseware which would flood the postwar market.

A replacement for urea formaldehyde came in 1937 with the rediscovery of a plastic called melamine, first discovered by a Swiss scientist in 1834. This plastic had mineral fillers offering better heat, moisture, acid, and abrasion resistance in addition to less color fading. Devine Foods, Inc. of Chicago was among the first to realize that these properties made melamine applicable to food storage containers. Since 1928, Devine Foods had been supplying hot lunches in metal containers to Chicago area factory workers. But because their metal containers had a tendency to rust, dent and sour food, the Devine Company began experimenting with melamine as a possible substitute for metal. Just as their prototype melamine food container was developed, America entered World War II. Melamine's first real trial then became a military one when the American Navy placed orders for unbreakable dishes that could withstand rough seas and gun fire impact as well as nest or stack efficiently in small ship galleys. Melamine met these requirements with ease, and the American military ordered millions of pounds of melamine throughout the war.

When the war ended, the American Cyanamid Corporation, the major producer of melamine in raw material powder form, sought a peacetime use for its product. Knowing it could create melamine powders in colors far more attractive than those used in grim military products, American Cyanamid set about proving the feasibility of melamine for a civilian market.

But companies buying melamine powder in the 1940s were using it primarily in familiar applications like special order institutional dishes, buttons or electrical casings. American Cyanamid's challenge was proving that melamine could have far greater profitability applied to domestic dinnerware. To do this, American Cyanamid set up a pilot factory in 1944 and commissioned prestigious designer Russel Wright to style the first prototype line of melamine dinnerware for experimental use in four New York City restaurants representing a wide range of serving styles from "counter service only" to "sit down" dining. Wright's successful trial design called Meladur proved melamine's viability, and within one year, the number of companies buying powder to mold into dishes jumped from two to eleven.

Russel Wright's initial experiment with plastic dinnerware was the thick-walled Meladur, first intended for restaurant use.

Melamine's potential was not lost on Wright himself who sold his rights to his initial rather pedestrian design, Meladur, in order to work on another, more stylish melamine line. Undoubtedly, Wright's early and continued interest in plastic helped lend dignity, credibility and seriousness to a material previously perceived as too untrustworthy and short-lived for dinnerware.

Like melamine, vinyl plastic had a similarly impressive impact on many areas of 1950s life. Throughout the decade, its use expanded far beyond its most predictable form, the shower curtain, into new areas such as drapery, window shades, tablecloths, luggage, rainwear, upholstery, convertible tops, and automotive interiors. In its challenge to both rubber and fabric, vinyl was perhaps the first plastic to seriously compete in applications not necessarily related to the kitchen and bath. But by comparison to melamine's impressive start, vinyl had a difficult beginning. Developed in 1925 from experiments at Union Carbide Company, vinyl made its appearance in consumer products in 1931. One of its first uses was in the creation of an unbreakable phonograph record, a function for which it was well suited. But it was only in the 1940s that vinyl technology advanced enough to produce heavier films suitable for flooring and upholstery. Flooring in particular seemed a natural application of vinyl, and a ready market was guaranteed by the large number of new tract houses being built on concrete slabs in need of floor coverings. Vinyl was both modern and gay but not as costly as wall-to-wall carpeting.

Yet early postwar products were disastrous. Although promising color stability and easy maintenance, the first vinyl floors faded and darkened when exposed to sunlight. Vinyl tiles also scratched and shrank creating crevices which filled with dirt. In houses without basements, moisture in the soil came up through concrete slabs and attacked tile adhesives.

Recovering from this shaky start was a major task; but considering vinyl's superior traits as a flooring material, its steady rise to acclaim over the decade is not so surprising. By 1954, better adhesives and coloring pigments had been found, and shrinkage and scratching were controlled. Vinyl finally proved it could be attractive without much upkeep. Its ability to function effectively in the hands of do-it-yourselfers undoubtedly assured its success.

Despite advances in vinyl technology and a large house buying public eager to purchase drapery, vinyl drapes remained largely relegated to an occasional perky appearance in the kitchen.

Vinyl as a drapery material had an initial launch perhaps even more disastrous than that of vinyl flooring. First introduced commercially in 1940, vinyl film achieved popularity in drapes in the late 1940s until the market collapsed in 1951 when consumers rejected the poor quality of most products. The first vinyl drapes were differentiated from shower curtains by adding printed designs which quickly washed or flaked away. Even more serious were technical problems that caused an offensive odor and stiffening a few months after purchase. In these early years, the vinyl film industry in general suffered from a lack of minimum standards and informational labeling. Often no indication was given on the product as to who made it or how to care for it. The problems in vinyl film almost succeeded in destroying the reputation of the entire plastics industry. And although these initial problems were corrected and a new embossing technology gave vinyl drapes a textured look, vinyl as a drapery material never mounted a sustained challenge to fabric.

But as an upholstery material, vinyl had remarkable success. The 1952 innovation of fabric-backed vinyl was no longer as hot, sticky or stiff as previous products had been. Consumers accepted it as a durable and easily cared for furniture covering that imitated luxurious, impractical and refined natural materials like expensive leather.

Vinyl's shower curtain image was lessened by connotations of elegance and luxury.

Vinyl on furniture allowed for the introduction of light colored upholstery, even on the most heavily used pieces.

*17*

**THE PLASTIC MATERIAL THAT TAILORS LIKE FABRIC**

Do-it-yourselfers could buy vinyl like cloth and "modernize" their old dining room sets.

Black iron and wire dinettes gave greater airiness and delicacy than the more ungainly chrome models.

The first all-vinyl car interior in the 1954 Chevrolet Delray Coupe indicated that vinyl had passed the most rigorous industrial testing, and its ability to have almost unlimited texture and color possibilities sparked consumer interest. Vinyl's impervious, easily-washed surface meant consumers could fearlessly purchase white or pastel colored furniture. Drab, workaday colors and textures that hid dirt, especially on sofa arms, were no longer necessary. Now even average families could have a white "leather-like" sofa, a powerful symbol of wealth and elegance even if it was plastic.

Like vinyl floor tiles, vinyl upholstery also gained from its ability to be used by the do-it-yourselfer, particularly a housewife reupholstering a detachable dinette seat (in "less than fifteen minutes" according to advertisers' claims). In the dinette field, plastic laminated table tops accompanied by chairs with matching vinyl upholstery were a huge hit. Appearing first as far back as 1934, chrome plated dinettes initially received a cool reception when metal was still deemed an inappropriate material for indoor furniture. But by 1951, chrome dinette sets were being sold at a rate of 125,000 to 150,000 per month, and eighty-five percent of these dinettes were veneered and upholstered with plastics. Throughout the decade, dinettes moved away from the original chrome legs and chunky lines to more sophisticated "contemporary" styles. But although the styling changed, laminated plastic and plastic vinyl remained essential components of the popular dinette.

Even celebrated designers like Raymond Loewy tried their hand at designing dinettes which soared in popularity starting in the early 1950s.

**18**

Plastic laminates became much more than utilitarian tools as their color and pattern selection expanded.

Chrome, formica and vinyl dinette sets had proved one very important point: plastics could succeed in furniture design, and the role of plastics in furniture quickly expanded during the 1950s. One example was the increasing use of plastic laminates on furniture. Although plastic laminates had been around since 1913 as industrial electrical insulation materials, their decorative potential and applicability to the consumer market was first recognized by the Formica Company which poised itself for a timely and profitable postwar expansion. Between 1947 and 1955, the Formica Company tripled its production.

By the early 1950s, formica laminate was already becoming firmly entrenched in the kitchen as counter and table top material. Old-fashioned enameled steel surfaces tended to chip, while formica had no such weakness and provided greater warmth and color potential than metal kitchen surfaces. As the kitchen area became more visible in the new 1950s open floor plans, materials like formica, which provided practicality plus "decorator colors" and patterns, were eagerly accepted. By 1953, a technological advance in pressure sensitive adhesives had made laminates even more popular as important do-it-yourself materials. Advertisments capitalized on this potential, urging husbands to fulfill their wives' "greatest dream"... the dream of new kitchen counters.

Prefabricated paneling, formed and coated with plastic resins, made remodeling rooms like this "Bunk House" into do-it-yourself projects.

Also popular with do-it-yourselfers was plywood paneling with its impervious laminated melamine surface. This affordable paneling was considered an effective and speedy remedy for difficult to repair cracked plaster walls in older homes. Advertisements lured the "Saturday afternoon mechanic" into embarking on do-it-yourself paneling projects with promises such as, "This Saturday - wood panel your dining area in time for Sunday dinner."

The particle board and plywood so popular for paneling and other projects, like built-in television walls and storage units, were made possible by thermosetting plastic resins which bonded wood by-products into workable sheet materials. These cheap wood sheets required no sanding, kiln-drying or planing and presented no difficult knots or twisting grain. Plastic resins made this now ubiquitous building material an affordable and versatile alternative to traditional wood products.

But it was Charles Eames' famous fiberglas reinforced polyester shell chair of 1949 that proved plastic could be much more than just a veneering material. During the late 1940s and early 1950s, extraordinarily strong and lightweight fiberglas reinforced polyester plastic (known as FRP) was heavily in demand by the military as America prepared for the Korean conflict and the cold war. This created a shortage in civilian applications of the plastic (such as boat building), applications already plagued by technical problems and the high cost of fabrication. But even after fiberglass reinforced polyester improved its technology, cost and availability, the furniture industry was hesitant to embrace the concept of plastic furniture. Plastic's mass production capabilities went against the long-standing craft traditions of the furniture industry. And although plastic shell chairs were popular in 1950s avante garde and "modern" interiors, plastics in furniture construction would have to wait for a fuller exploration in the 1960s and 1970s.

Eames used industrial materials and wartime technology for his solution to the problem of an affordable, well-designed, modern chair.

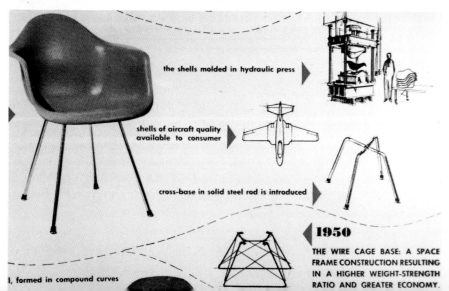

the shells molded in hydraulic press

shells of aircraft quality available to consumer

cross-base in solid steel rod is introduced

1950

THE WIRE CAGE BASE: A SPACE FRAME CONSTRUCTION RESULTING IN A HIGHER WEIGHT-STRENGTH RATIO AND GREATER ECONOMY.

l, formed in compound curves

Both vinyl and fiberglas reinforced polyester met with mixed results in their attempt to enter the 1950s living room. But two other wartime plastics, polystyrene and polyethylene, achieved phenomenal success in the 1950s by staying right where plastic had previously found a home - in the kitchen.

These two post-World War II kitchen plastics benefited from precedents set in the 1930s by the colorful, lightweight, durable plastic premiums given away by Depression era companies facing poor sales and highly competitive markets. Cheerful plastic premiums proved a tantalizing incentive for consumer buying, and the volume of premium manufacturing doubled between 1930 and 1941 as plastics entered millions of American homes. (The General Mills Corporation alone distributed nearly 30,000,000 plastic premiums during this time period.)

Prewar premium items such as bathroom cups, cereal bowls, salad forks, coffee spoons, cookie cutters, and cleanser holders made of urea formaldehyde or cellulose acetate were so popular that they were later introduced as plastic retail items in chain stores after their premium status had elapsed. Wartime consumers continued this enthusiasm by eagerly snatching up the few plastic kitchen items made from scraps not being used in the war effort. The postwar plastics, polyethylene and polystyrene, undoubtedly benefited from this tradition of support for plastic kitchenware.

The first widely used kitchenware plastic was polystyrene, a thermoplastic material characterized by a clear, glasslike, high gloss surface. Polystyrene was first introduced in 1939, although its development for consumer product use was interrupted by World War II. Polystyrene was in plentiful supply after the war, however, since it had been the chief ingredient in the manufacture of wartime synthetic rubber, and its production was quickly resumed. As a cheap, speedily injection molded material, it proved ideal for inexpensive kitchenware. The Federal Tool Corporation of Chicago, for example, was stamping out 700 compartmentalized picnic plates per hour in 1952 with virtually no waste of materials. Polystyrene rejects or scraps were simply reground for use in the next batch of plates, an easy and extremely effective use of materials.

In addition to all of these circumstances favorable to polystyrene's growth was the wartime scarcity of metals traditionally used for kitchenware like bowls, trays, measuring cups, canisters, and bread boxes. With all its attractive characteristics - color, luster, clarity, low cost, resistance to acids and alkalis and some solvents - polystyrene had few rivals and was the dominant kitchenware plastic until the late 1950s.

Polystyrene Wata-Kanta Pitcher from Plastic Products Company of Los Angeles, California

Thermoplastics like polystyrene and polyethylene were used for dishes, but proved less durable than the stronger thermosetting plastic, melamine.

Polystyrene made cheerful, sleek designs a part of routine daily chores.

Some kitchen plastics were manufactured in more subdued colors for those who preferred a less flamboyant kitchen.

This shaker set from Plas-tex illustrates the 1950s dome shape, well-received for its simple and easily cleaned lines. In contrast, the Lustro-ware napkin holder demonstrates plastic's ability to create intricate decorative forms as well.

Polyethylene's waxy toughness outperformed even the highly successful polystyrene.

Every day's a picnic with new

CAREFREE

WORK-EASERS

Housewares by **Plastray Corporation.**

The chief competitor of polystyrene plastic was polyethylene, developed in England and introduced in the United States in 1943 as cable insulation for radar in World War II. Unlike the cool, glossy surface of polystyrene, polyethylene was a soft, flexible, waxy thermoplastic, most widely known for its use by chemist Earl S. Tupper in his fabulously successful Tupperware line of household products. Like polystyrene, polyethylene (sometimes referred to by its duPont Company trade name, Polythene) was resistant to chemicals, highly colorful, tasteless, and odorless. And as the lightest of all commercial plastics, it was the only one light enough to float on water. These traits made polyethylene immediately applicable to food and beverage storage containers with squeezable pouring spouts or tight sealing lid covers like those Tupper made famous. One of the most miraculous traits of polyethylene was its ability to withstand the temperature of a freezer without crazing, cracking or becoming brittle.

By 1953, scientists had discovered a way to control the order of the molecules which composed polyethylene. With this breakthrough, polyethylene could be designed as a stronger, tougher and even boilable material. Polyethylene - with all the benefits of both its new and its older forms - now truly dominated the plastics industry. By 1958, polyethylene had eclipsed polystyrene in the kitchenware field by expanding beyond small ware into larger houseware such as trash cans, pails, laundry baskets, and dish pans. As an undentable, noiseless and non-abrasive material, polyethylene surpassed both metal and polystyrene in production and sales of kitchenware.

### Tupperware

In 1945, chemist Earl S. Tupper began the manufacture of a line of polyethylene kitchen products. Sensitive to the bad reputation of plastics during the late war years, Tupper slyly proclaimed that his products were made of "Poly-T" not plastic commenting, "There have been too many bum articles called plastic." The response from the design establishment to the Tupperware products was immediate. The Museum of Modern Art included them in its show of well-designed useful objects and *House Beautiful* of 1947 proclaimed them to be "Fine Art for 39 cents," reminiscent of "alabaster and mother of pearl with a profile as good as a piece of sculpture and the fingering qualities of jade"!

Tupperware originally appeared in twenty-five translucent, frosted pastel shades in a variety of kitchen container items, all having Tupper's specially patented airtight seal. Knowing that this seal was what would separate Tupperware from all other future polyethylene rivals, Tupper instituted special store demonstrations of the seal for potential buyers. But in 1951, unhappy with retail demonstrations of the seal, Tupper removed all of his products from retail outlets in favor of a "home party" sales force patterned after the success of companies like Stanley Home Products.

In three years time, sales had tripled; and the sales force had grown from 200 to 9,000, ninety-five percent of them housewives. With a motto of "Build people and they will build the business for you," Tupperware Home Party Inc. gave its housewife dealers a sense of purpose and recognition for an achievement which, although confined to the "feminine sphere" of home, was rewarded with some of the money and sometimes even the cars, impressive clothes and travel usually available only to men.

The famed Tupperware vacuum seal meant containers could go into the refrigerator sideways or even upside down without spilling.

Tiny flecks of gray and a "silver" domed knob enhance this otherwise workaday polyethylene pitcher.

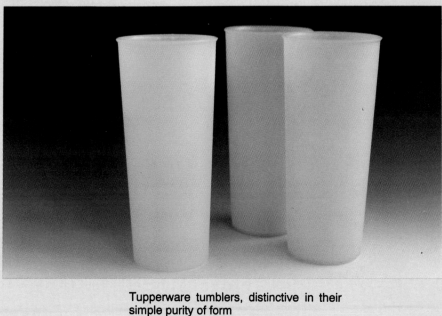

Tupperware tumblers, distinctive in their simple purity of form

Regal as the splendor of imperial Rome...

VINA-LUX® FLOORS ARE *Chic*

Plastics often overcame consumer suspicion by masquerading as expensive luxury materials or associating themselves with images of elite elegance.

Chairs...care-free'd by **V I N Y L**

"Micarta is ideal for all fine furniture"

*Edward J. Wormley*

EDWARD J. WORMLEY IS A PROMINENT DESIGNER OF HOME FURNISHINGS AND INTERIORS.

# Chapter 3:
## A Housewife's Dream

It is difficult to believe that there was ever a time when it seemed quite possible that postwar plastics would be doomed by lack of consumer acceptance. Indeed plastics promoters had much work to do to overcome the disasters of the 1946 flood of shoddy products that reached a naive wartime market starved for consumer goods. But because there were so many technological and production breakthroughs occurring throughout the decade, the industry benefited from an indomitable spirit of innovation and excitement as did consumers who were swept up in a flood of new plastics and plastic products.

Marketing strategists intent on creating a consumer preference for plastic had a rich choice of possible approaches open to them. One of the safest and most successful options, particularly in the early years of postwar plastics was to stress one of plastic's oldest traits: its ability to imitate a variety of expensive, traditional natural materials. This was an old and familiar way to think of plastic, and a bit of plastics history repeated itself. If Victorian plastics had achieved popularity based on an ability to imitate expensive natural materials like tortoiseshell and ivory, perhaps a similar strategy would work for 1950s plastics.

In this spirit, ads claimed plastic dishes had the "lustrous hard glaze of porcelain and the weight of china;" plastic tile looked like "smart expensive ceramic tile;" new, indestructible vinyl wallpapers could create "the delicate watercolor beauty of the finest wallpapers;" plastic curtains would look just like "delicate and lacy" sheers yet never need ironing or more laundering than a "quick wipe with a damp sponge;" scrubbable, impervious naugahyde plastic upholstery could simulate leather, tweed, tortoiseshell, damask, bamboo, bark, wicker, or rope. (Naugahyde's "English Manor" and "Burnished Antique" upholstery reached for obvious connotations of wealth and tradition.) Even plastic laminate formica was promoted as looking just like "any fine hand-rubbed furniture finish."

In this selling strategy, plastic became a democratizing influence bringing luxury, prestige and tradition within affordable reach of average consumers. But although this selling technique was successful, it also presented a problem for the plastics industry. Stressing plastic's unique talent for imitation perpetuated stereotypes of plastics as "mere substitutes," clever but cheap and inferior. Implied in this type of advertising was the unintended message that as soon as one could afford it, one might better buy "the real thing."

In an attempt to overcome this prejudice and prevent a consumer shift back to natural materials, advertisers had to prove plastic had other special abilities. So an appeal to the consumer desire for luxury was often coupled with an equally strong appeal to old-fashioned common sense. Wasn't it "obvious" that tough and easily maintained plastic "luxury" was better than delicate, traditional materials which required special time consuming care? This argument was used with great success in an era which centered on young children and a relaxed entertaining style.

Plastic laminates and upholstery in particular benefited from this type of marketing since "cultured" wood, faux marble and pretend leather could promise beauty plus practical convenience. Consumers had seen vinyl used in upholstery for buses, autos, trains, theaters, and hotels. They had seen lobbies, libraries and restaurants with laminates on table surfaces. But to bring these commercially used materials into

At times, the effrontery of plastic molders seems boundless as in these kitsch imitations of Wedgwood and Belleek.

Imaginative modern forms were possible when plastic was used adventurously rather than imitatively.

a domestic setting, consumers had to perceive them as imbued with connotations of beauty, luxury and "tradition" more suitable to the private home. Modern science had indeed brought undreamed of practical benefits which plastic's marketers skillfully mixed with a comforting dose of luxury and tradition. Otherwise postwar plastics might have found themselves too serviceable for their own good.

To target more adventurous consumers, plastics could be equally well presented as embodying the very spirit of progress. Their colors could "fit today's palette" and their forms could be daringly organic, asymmetrical, smooth, and sleek - enough to please even the most avante garde buyers. Presented in this light, plastics seemed to be breaking free of their "cheap substitute" image. Accepted as materials in their own right, plastics' special traits when rightly understood, could inspire the complete redesign of old products. And a small, elite group of designers and design-conscious consumers did indeed embrace plastics in furniture, lighting and tableware - areas where plastics had never been seriously applied.

But in those areas where plastics had never before succeeded, many obstacles had to be overcome even among the most receptive buyers. For instance, convincing consumers that plastics were suitable for everyday dishes, not just picnic dishes, was perhaps the most formidable marketing challenge in the industry.

Consumer suspicion and skepticism were hardly surprising considering the poor performance of urea formaldehyde dishes. Even more daunting was the problem of how to overcome resistance to melamine's price which was often higher than earthenware and some china. In 1953, the lowest price melamine sixteen piece dish set was $9.95 while a sixteen piece china set could be purchased for only $5.95.

It also must have seemed apparent that melamine's break resistance would ultimately inhibit growth and profit since theoretically, no dish replacements would ever be needed. And there remained the problem of melamine's decorative limitations. Before 1954, it could only be molded in a limited range of solid opaque colors, which seemed far less expressive than the wide variety of patterns and colors available in ceramic dinnerware.

The marketing campaign for plastic dinnerware developed by the American Cyanamid Corporation to overcome these problems is representative of the tactics used by many other plastics industry leaders who saw the need for a combination of industry-wide high performance standards and better labeling information to educate consumers and retailers on what to expect from and how to care for each particular plastic they purchased.

Starting in 1949, American Cyanamid orchestrated an all out campaign to reach consumers, retailers and dish buying institutions. To reach the institutional market, manufacturers distributed informational booklets to home economists and managers of facilities like hospitals, cafeterias and restaurants serving food to large numbers of people. For consumers and retailers, ads and articles on the use and care of melamine were placed in home decorating magazines and trade journals. Department stores received kits suggesting how their local ad campaigns, radio commercials and window displays could be tied to the industry's national campaign. Salespeople could view a ten minute training film called "The Bull in the Melmac Shop," and in some cases, trainers were hired to monitor and assist melamine's arrival at various stores.

But effective labeling was really the single most crucial element of melamine's marketing plan. Devised with care and precision, labels had to create demand by guaranteeing that plasticware had at least one thing ceramicware lacked - unbreakability. Products labeled "Melmac" (American Cyanamid's trade name for melamine) were promoted as meeting the highest standards of unbreakability. This advertising strategy trained consumers to look for and trust the Melmac label so that no

*He lost the load, but not the dishes!*

Restaurants with high dish breakage costs seemed a promising market for plastic dishes, although this ultimately did not come about.

Compact packaging and colorful displays attracted consumer interest in melamine dinnerware.

To demonstrate melamine's durability, this department store window featured a running washing machine filled with plastic dishes.

Plastic industry leaders hoped that newly educated consumers would boycott sloppy manufacturers and ignorant salespeople.

matter which particular firm was molding the dishes, buyers would still know that the raw material was of the highest quality.

Putting the consumer and retailer at ease was of particular importance in these early days of the new postwar plastics. The speed at which the plastics industry was growing in these early years produced a bewildering number of new plastics and new companies, each with its own trade name for its material. The late 1940s with its melting dish drainers, cracked food containers, wary consumers, and futurama hyperbole, meant that educating the public and retailers about the capabilities and care of various plastics was crucial if the industry was to thrive in the 1950s. But with devices like the Melmac label, which educated buyers and sellers on melamine's traits and care, the plastics industry was successfully able to address its credibility problems (not to mention receiving the obvious benefits of brand name recognition and repeat sales).

Labeling products increased consumer trust by turning plastic's bewildering complexities into readily identifiable brand name loyalties.

In the case of melamine dishes, even more credibility was gained when melamine producers joined with the U.S. Bureau of Standards to arrive at both commercial and domestic voluntary standards for the industry. Thanks to an energetic and expensive Cyanamid campaign aimed at both retailers and consumers, melamine avoided the potential for confusion and distrust. By 1952, eighty-five percent of the country's two hundred leading department stores carried at least one melamine line.

But like any other product, quality plastic was not cheap even though the public may have expected anything made of plastic to be automatically inexpensive. In the case of melamine, expense was justified by stressing durability. As one pleased housewife told market researchers in 1949, "Your initial cost is your final cost." Consumers had to be convinced that by spending just a little bit more, their savings would increase dramatically in the long term. Manufacturers could only hope that the child-centered world of the baby boom 1950s would help launch their child proof product and that its impressive efficiency would sustain future sales.

From the molder's point of view, packing costs for an unbreakable product obviously amounted to little; and with plastic dishes in particular, the elimination of unpredictable kiln drying meant there was none of the expensive rejects inherent in ceramic production. These extra savings could be spent on ingenious department store displays and convincing warranty labeling.

New outlets for plastic dishes, including small gift stores, grocery stores and door-to-door sales, were made possible by clever breakthroughs in packaging design. Chicago Molding Products, for example, sent dinnerware samples to these new outlets in a special "three in one" pack. A single box served as a shipping container, a carrying case (with pop-up handle good for door to door sales) and a self-contained display unit that simply slid out ready for inspection. This packaging was far more efficient than that used for most china which arrived at stores in straw packed barrels that had to be unpacked, sorted and then packed again into ready to take home sets for the consumer.

How costly are your flying saucers?

Consumers were urged to consider the long term expense of replacing broken china when deciding whether to invest in plastics.

Green or Pewter Gray. Manufactured by the Boonton Molding Co., Boonton, N. J.

Very Special—for Christmas!

Beautiful Boontonware

16-piece dinner service

with special gift Salt and

Pepper Set . . . only $14.95†

(value $18.75)

†Slightly higher in the West

Boontonware®

*WRITTEN GUARANTEE WITH EVERY PURCHASE

BIG SAVINGS ON BIG SETS—SAVE $8.00 ON COMPLETE SETS FOR 6 OR 8!

Ingenious packaging made plastic products appealingly convenient to buy and to sell.

29

The ingenious, compact packaging designed for melamine also encouraged consumers to change their buying patterns. The prospective dinnerware buyer was told that sacrificing and saving for a huge matched set of china dishes was old-fashioned. Instead designers Russel and Mary Wright came up with their new marketing concept of the "starter set" which enabled smaller investments in dishes and a more irreverent attitude toward their use.

With this aggressive marketing strategy, melamine experienced remarkable success. From 1942 to 1960, the number of companies molding melamine into dishes jumped from two to twenty-four and production of melamine powder grew from fifteen million pounds in 1953 to fifty-five million pounds by 1959.

Other plastics manufacturers lost no time in following similar marketing strategies. Dow Chemical, maker of polystyrene powder, was among the first to realize that consumers were put off by the complicated terminology of the plastics industry and would likely find reassurance in trademark recognition. Its national public relations campaign harnessed the selling power of a simplified trade name for polystyrene, "Styron" which became a name consumers could recognize and trust as a guarantee of good quality plastic provided by Dow to the various companies who molded kitchenware.

Right: Kitchenware sets expanded as rapidly as their "round-the-clock" usefulness.

Heavy and awkward in shape, "portable" TVs were more cumbersome than "portable," although depictions such as this one made them seem light.

To increase sales even further and take advantage of consumer enthusiasm and name recognition, many kitchenware molders expanded traditional four canister kitchen sets to include such items as a matching bread box, cleanser holder, towel dispenser, cookie jar, wastebasket, garbage can, tray, and picnic set. The packaging of kitchenware also became increasingly sophisticated as product quality increased, and the gift buyer market was targeted. Even Emily Post advised, "For the first anniversary - colorful, convenient, practical plastics along with paper, make the delightfully modern anniversary gift."

As interest in plastics grew, national chain variety stores began to display plastic kitchenware in mass. The striking colors and sparkling finish of these plastic products grouped in an impressive array created a stir among shoppers. Entrancing, vivid colors became a way to "stop the shopper, then close the sale." The success of plastics displayed in this manner sparked the interest of independent retailers and department store buyers who soon began to stock plastic kitchenware in their own stores, utilizing plastics' color impact as a marketing tool. Supermarkets too seized on plastic kitchenware as highly marketable "tie in" items (for example, plastic flour sifters sold next to cake flour) with a higher margin of profit than regular foodstuff items.

**30**

The "patio portable" concept drew on the popularity of TV viewing and outdoor living.

At times, these marketing campaigns had unexpected results. The Bolta Company's vinyl upholstery, Boltaflex, was among the many vinyl upholsteries that experienced initial resistance from furniture manufacturers and retailers who were particularly entrenched in traditional manufacturing methods and materials. These makers and buyers endorsed only leather and fabric furniture coverings, not an unknown product called "Boltaflex."

The Bolta Company successfully changed these attitudes with a combination of product tag labeling, direct mailings, trade magazine ads, and Boltaflex swatch books consumers could ask to see. So effective was the consumer ad campaign that it soon became a case of the "tail wagging the dog" - furniture was being sold as much for the appeal of its upholstery as for its craftsmanship. Manufacturers and retailers were pushed toward Boltaflex because consumers were asking for it by name.

In the case of the television industry, plastic became an important material in the marketing sensation, the "portable" TV introduced in 1955. But despite ads which showed smiling women carrying these televisions in one hand, portables weighed an ungainly thirty to fifty pounds and often had uncomfortable handles. Plastic cabinets helped somewhat in reducing weight but until 1960 portables were squarish and boxy in design, making them difficult to comfortably carry next to the body as one would a suitcase.

Despite these problems, portables were very popular and helped rescue the struggling TV industry which, by 1954, had managed to sell at least one set to nine out of ten American families. Facing a glutted market and a spiralling decline in profit margins, the TV industry struggled to create fresh demand by promoting the concept of the portable and fun "second set." By 1958, portables accounted for thirty percent of the televisions sold as consumers found their lower cost and fairly adequate picture quality appealing. This strategy turned sour though when consumers began buying cheaper portables as their primary TV set, content to let the less expensive and still heavy fourteen to seventeen inch screen sit permanently on a a table or divider book shelf. The industry's dream scenario of selling one expensive first set to every family and one portable to each family member seemed destined to fail. High demand and high profits would have to wait until the next breakthrough, color television.

Televisions in fashionable "two-tone" color schemes were made possible with plastics, as in Sylvania's Dualette of 1959.

**31**

ARVIN Your choice of three cabinet colors – green, ivory or sandalwood.

CAPEHART Cabinet finishes cover a range of six smart colors.

FIRESTONE "Slumbertone." Utility model; white cabinet, luminous dial.

JEWEL Ebony or walnut with ivory grille, or ivory with maroon grille.

MOTOROLA Cabinets in the always popular tan, green and ivory shades.

PHILCO Cabinets available in mahogany, white or ivory finish.

STROMBERG-CARLSON "MusiClock." Ebony and red; silver area trim.

TRAV-LER Hand-rubbed wood cabinet in mahogany or blonde finish.

Colorful and cheaply produced plastic housings helped make 1950s radios into disposable "accessories."

The most adventurous technological experiment to emerge from this period was the unusual 1959 Philco Predicta, a design which attempted to revive consumer interest with a startling new "look." Philco designers encased the TV picture tube in plastic and removed it from its cabinet. Mounted on top of but visually separate from its console, the Philco's picture tube swivelled to accomodate different viewing positions. A portable model had a twenty-five foot cord, allowing viewers to take just the view screen to another room in the house. Ads declared, "Imagine! Move your picture from room to room without moving the set!" Unfortunately, with the controls located in the other room, adjusting the picture proved impossible and the twenty-five feet of plastic cord was both hazardous and unsightly. Even without its control panel, the picture tube was just as uncomfortably heavy to carry as the other "portables." Nevertheless, this innovative use of plastic housing helped the television manufacturing industry sustain the marketing image of TVs as ever new and different even when these changes were superficial or gimmicky at best.

The radio industry faced the specter of declining market demand a decade before television makers. By 1947, ninety-three percent of American homes had at least one radio. Demand was falling, and many feared the radio industry would certainly begin a severe decline even though television had not caused its extinction as some had predicted. But the new thermoplastics of the 1950s continued a tradition begun in the 1920s of durable plastic radio cases in bright fashion colors. Cheap to produce and available in an ever widening number of color possiblities, plastics made it possible for radios to be "restyled" in an almost endless variety of forms.

The industry's goal of a radio in every room and a radio for every family member seemed at last within reach with the inexpensive styling changes made possible by cheap thermoplastics. And if altering the style of cabinets didn't sufficiently stimulate sales, radio manufacturers added new technical "improvements" such as the "clock radio" or "midget radio," both 1950s inventions. Some radios were given "special features" like having handles that "disappeared" or doubled as stands. Others were designed to appeal to a specific market like teens, children or women. ("Feminine" radios for example masqueraded as cosmetic cases.) These inventive sales tactics in low cost 1950s radios were often made possible by a shrewd use of plastic.

Light thermoplastics and transistor technology enabled radios to shrink to the size of a postcard and weigh less than a pound.

The gadgetry of clock radios suited the 1950s notion that modern technology was like having many talented but invisible servants.

The long, low lines of radios echoed trends in 1950s sofa design.

### Luggage

Luggage sales skyrocketed after World War II as Americans increasingly enjoyed motor vacations and air travel. Once viewed as a "slow seller" with a maximum of two or three sales over an average consumer's lifetime, luggage now became another heavily marketed "fashion" item whose features were touted as ever more "modern" and miraculous.

Ads tried to convince consumers that they needed the "right" set of luggage for every type of trip or at the very least, all the right pieces within one set. A "luggage wardrobe" might have a pullman case, a train case, a weekender case, three different sized suit bags for men, duffle bags for miscellaneous small items, hat boxes, and cosmetic bags. And since luggage was marketed as part of a "fashion" ensemble, men's and women's baggage came in "feminine" and "masculine" colors and shapes.

Plastics, particularly vinyl and fiberglas reinforced plastic, had much to do with this new and highly successful marketing strategy for luggage. By 1954, plastic was even replacing plywood as the major structural conponent of luggage. Fiberglas reinforced plastic technology had advanced enough to make it cost effective for manufacturers to experiment with this new light but extremely strong plastic which could be covered in vinyl or left plain to reveal the glass fibers imbedded in the plastic. The streamlined shapes obtainable by FRP plastic were declared "as sleek as the plane they fly in."

Protective, impervious vinyl sheeting or vinyl coated fabric proved superior to canvas and leather as a luggage covering. In 1948, Shwayder Brothers, makers of Samsonite, came out with the first all plastic covered suitcase with even its hardware made of molded vinyl. Plastic was simply cheaper, tougher and more colorful than leather in particular, whose market share of the retail luggage industry had dropped to twenty percent by 1956.

Vinyl suitcases could look like leather yet resist the scratching, staining and scuffing leather baggage often suffered.

A BOLD NEW CONCEPT IN LUGGAGE

The strength and light weight of FRP plastic had a significant impact on luggage construction in the 1950s.

for the young in heart

New developments in plastics helped fuel the 1950s do-it-yourself movement.

Little tables, the versatile work horses of 1950s decor, were frequently given a protective plastic surface.

With general consumer resistance lessening by mid-decade, the plastics industry turned its attention toward a new, fast growing specialty market, the do-it-yourself Saturday handyman. Factory automation and work saving home appliances had created increased leisure time, and many 1950s families used this new leisure time to experiment with affordable home improvement weekend projects. The tremendous popularity of TV viewing also meant more and more Americans were spending evenings at home, and the desire to "fix things up" around the house became strong. Power tools and paint sales boomed. But plastics were not to be left behind in the rush to capitalize on the do-it-yourself fad.

Because plastic products could be applied with fewer tools, less skill and a smaller investment than many other materials, they seemed perfect for an era of budget-conscious young families and high labor costs. Any lingering doubts about the social stigma of using an inexpensive "substitute" material were quelled by decorating magazines which often reassured young homeowners that "imagination" counted for more than money in creating a home environment.

Most plastic do-it-yourself products involved sheets of laminates that were easy to handle and required little trimming or sanding and no painting or staining. Advances in adhesives brought still more popularity to plastics that permanently bonded to a surface with only the pressure of a roller. Laminates even began to be packaged in rolls that a handyman could carry out to his car.

A host of projects beckoned the ambitious plastics handyman: old furniture updated by colorful impervious laminates, vinyl floors applied in do-it-yourself kits, plastic lawn chair webbing substituted for old faded canvas, and dinette chairs recovered in the latest vinyl upholstery. Uneven cracked plaster walls "disappeared" under plywood laminated paneling or polystyrene tiles, and plastic "luminous ceilings" covered ugly bare bulbs in entranceways. Plastic corrugated panels were especially popular for creating carport walls, patio roofs, terrace screens, entrance canopies, even interior divider walls.

Enthusiastic amateurs chose plastics for quick and professional looking results. In response, plastics manufacturers increased availability by distributing plastic materials at a wider variety of outlets. Lumber yards, department stores, paint stores, floor dealers, even appliance shops - all became places where a do-it-yourselfer could purchase plastic materials.

And for the wary or inexperienced handyman, newspaper, radio and store counter displays advertised pre-assembled kits, decorator idea booklets and instruction sheets. The do-it-yourselfer of the 1950s looked to plastics as one way to modernize an old home or add distinctiveness and individuality to a tract house or apartment. Efforts to market plastics to this group of consumers found an audience ready to put plastics to work with their own hands.

34

Vinyl wall sheeting emulated the "rich look of old brick" or the "pleasant roughness of stone" - elements of time and nature missing from modern pre-fabricated houses.

Translucent corrugated plastic panels transmitted light but ensured privacy without creating walled-in cramped spaces.

China companies had traditionally been able to attract consumers by depicting dish selection as a dreamy, romantic interlude.

# Chapter 4:
# Tradition Fights Back

As the 1950s progressed, the success of plastics in home furnishings began to present a serious threat to traditional materials. Would plastic make materials such as wood, metal, leather, and china obsolete? Some manufacturers fought back to ensure their survival. At times though, it was a losing battle. In the kitchen in particular, polystyrene and polyethylene achieved such immense consumer popularity that metal and earthenware had little hope of regaining their dominant position in the kitchen accessory market. But not all of plastic's competitors would acknowledge plastic's superiority. The battle which occurred in the dinnerware field between the new melamine material and traditional china and pottery was particularly fierce.

The china and earthenware industries, already struggling against cheap imports, looked on with dismay at the melamine "invasion." For although the ceramics industry continued to grow modestly during this period, imports had skyrocketed. Between 1947 and 1956, imports of china and tableware, many from Japan, increased by almost eight-hundred percent while imports of earthenware nearly doubled. Melamine's sudden popularity further eroded china's sales base, causing a twenty-five percent loss of its former market share. It seemed clear that the ceramics industry had to take some form of action.

In 1953, the Vitreous China Association attacked plastic dinnerware as a spreader of "contagion and disease" among the American people. Scientists hired by the china industry claimed that easily made knife cuts in melamine's surface would become "harboring places" for bacteria never fully removed even by machine dish washing. The plastics industry countered that normal dish washing eliminated bacteria and that the so called knife cuts were actually only "indentations" that occurred "under considerable pressure." (This accusation that melamine scratched was partially neutralized by proof that china could and did scratch also, particularly when the unglazed foot of a dish marred the surface of the dish beneath it.)

Both industries brought forward conflicting evidence as to plastic's thermal conductivity. China producers, whose plates needed to be warmed in an oven to keep foods hot, denied that melamine's low thermal conductivity meant it was able to keep foods hotter or colder more efficiently than ceramics. Disputes such as this went forward inconclusively as each industry presented its own scientific "evidence."

To plastic's defenders, the most outrageous claim made by the china industry was that melamine released formaldehyde into warm or hot liquids. Denying the implication that using melamine was like drinking embalming fluid with your coffee, an angry industry spokesperson sarcastically commented, "Soon no doubt the citizens of this country will have an epidemic of plasticitis and the undertaker won't be necessary because they will already be embalmed." But despite all its accusations, even the American Ceramics Society had to admit that melamine dinnerware had a "surprising number of advocates."

The ceramics industry was particularly hampered by an advertising tradition that suggested prim formality and a distribution system slow to use pre-packaged starter sets. Some manufacturers sought to change their image with new ad campaigns stressing modernity and informality. Gladding McBean of California, makers of Franciscan ware, launched a new line of middleweight china to compete directly with melamine. Others simply waited for consumers to return to their senses and to ceramics as an obviously superior material for dinnerware. Some ce-

The streamlined packaging of compact melamine starter sets was in marked contrast to the sprawling elegance of a full china set on display.

The diversity available in melamine cup handles typifies the quick design changes and innovative experiments in form that more tradition-bound china patterns could not match.

Iroquois China engulfed its dishes in flames to point out their superior heat resistance compared to melamine products (although thoughtful consumers might have realized such a fiery scenario was both unlikely and inadvisable).

By changing their focus from engaged couples to young families, some china makers tried to slow down the shift toward plastic.

The boxy, wood TV console did not fit particularly well into modern decor.

The TV age seemed to call for an end to the absurdly irrelevant antique pretensions satirized here.

ramics companies, such as Lenox and Royal China, entered the plastics field themselves by the late 1950s since there seemed no denying that melamine just was not going to go away.

In 1957, *Ceramics Industry* magazine reluctantly acknowledged the trend toward plastics by interviewing an anonymous ceramic manufacturer about to open a plastics division. Admitting that ceramic dinnerware was faltering, the manufacturer tried to pinpoint the cause: "Ceramic dinnerware advertising is tied with tradition, beauty, dreams, heirlooms and many other ho-hum types of themes...They [plastics ads] appeal to the practical, modern, fast moving mind - in short, they have zip."

In other fields, plastics were received with a similar pragmatic attitude. The special popularity of portable TVs, radios and phonographs meant that cheap, colorful, lightweight thermoplastics could readily compete with more traditional cabinet materials such as wood or the older and more expensive pre-World War II phenolic plastics. But some TV makers pointedly advertised their products as "genuine wood - not plastic," promoting the idea that TVs should blend into room decor disguised in fine wood cabinetry.

This appeal to luxury and tradition did not, however, meet the needs of informal 1950s living where the light, vibrantly colored portable TV set was a desirable object. An enormous three story, 2,000 ton hydraulic press was needed by Chicago Molded Products Corporation to mold the first phenolic plastic TV cabinet in 1950. Despite the obvious difficulties of using such gargantuan tooling, this first molded plastic cabinet showed the kind of production speed, lack of finishing costs and overall durability that plastic products and production methods could deliver to the TV market. By using this plastic cabinet, Admiral Television Company was able to price its product one-hundred dollars cheaper than its nearest competitor's product, a significant cost reduction indeed in what was a highly competitive market. And even with its own timber stands, lumber mills and cabinet factories, the Philco Company found plastic to be fifteen percent cheaper to use as a TV cabinet material in a 1950 study of the issue.

Whether brought about by a wheeled iron cart, a swiveling picture tube or a simple carrying handle, portability in TVs expressed the mobility of the "space age" 1950s.

Plastic was particularly suited for the portable TV because without the visible joints, nails, and screws of wood construction, the molded plastic cabinet could look good from all sides, a trait which was to become increasingly important as the TV was made to sit on a table in the middle of a room, visible from many angles.

Advances in picture tube technology were also making it possible to enjoy larger and larger TV screens. For these larger, non-portable cabinets, using sheet metal laminated with a "wood grained" vinyl proved more economical than the all plastic cabinet. But whether one wanted a traditional large cabinet or a fun portable, plastics were central to the appearance of both. The dual talent of plastic to simulate wood grain or assert its own cheery, colorful light weight made fighting its position in the television market economically and aesthetically unwise.

The leather industry also found itself faced with a plastics challenge. Indeed by 1952, leather makers began to realize consumers would no longer automatically accept the superiority of what was traditional in upholstery. Vinyl upholstery had not fallen out of favor with postwar consumers, even when leather was once more available after the war. Shortages and uneven quality of animal hides continued to plague the leather industry while vinyl upholstery makers manufactured with complete uniformity and low cost in patterns that imitated leather yet seemed to consumers to "improve" on it as well.

With leather sales slipping, manufacturers banded together to start a national campaign which would promote leather without engaging in the direct attacks on plastic which had characterized the dinnerware battle. By emphasizing the luxury and quality appeal of their product, leather makers played on the consumer's desire for status, implying that vinyl was a mere stop gap material for those hectic, low budget first years of marriage and parenting.

In a 1953 *New York Times* ad, leather makers reminded retailers that plastic upholstery would not long satisfy the changing desires of 1950s families. The ad announced: "Jonesism is Dead! Nobody wants to keep up with them any more. Everybody wants to be different from the Joneses. That's the big news of 1953, Mr. Retailer... Are you alert, Mr. Retailer, to this ever mounting pressure of Mr. and Mrs. America for distinctiveness, individuality, quality...There is nothing that can so meet this demand in your business so simply, so naturally as leather..."

Eventually, china makers too would regain their composure and turn back to ad campaigns that stressed the status appeal of traditional material. To beat their plastic rivals, both industries labeled their competition mere "substitutes," an effective use of that oldest of all prejudices against plastic.

Sleek plastic encased the mechanical backs of televisions making them tidier to look at and easier to position.

Vinyl upholstery gave more color choice and better wear resistance than leather products.

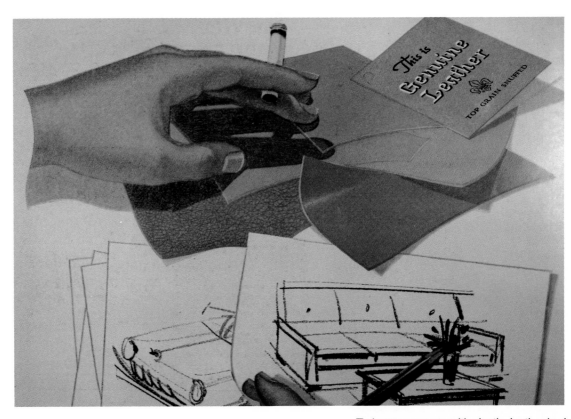

To better compete with plastic, leather had to be repositioned in the marketplace as the material of choice for progressive designers.

With the war at last over, young Americans rushed to begin marriage, family, career - and consumer spending.

No matter what trouble tots got into, plastic dishes promised to emerge unscathed.

"Darling! let's give truly useful gifts this year!"
Dinnerware molded of **MELMAC**®
stays handsome, helpful and whole

What to buy the numerous couples setting up households in the 1950s? Why, practical plastic dishes, of course.

Plastic products, such as these portable phonographs, were presented as wise insurance against family disintegration and generational friction.

# Chapter 5:
## When Their Halos Slip...
## The Child Proof Home

Plastic's remarkable success was due, in part, to its ability to meet the physical needs of the fastest growing socio-economic group of the 1950s, young families. With limited budgets and accident-prone toddlers, many turned to plastic in the home as a "miraculous" boon both to the family bank account and the family morale. For although plastic materials sometimes required a larger initial investment, their higher cost was justified by their great strength and durability. Extensive china patterns of extreme fragility and formality, for example, may have seemed both extravagant and impractical for young couples with young children. And as the postwar marriage and birth rates soared, wedding and anniversary gift buyers may have considered starter sets of melamine dishes to be wiser and longer lasting gifts than china.

The popular plastics wisdom of the day promised that the indestructibility of plastics diminished the likelihood that "Junior" would suffer from what one advice columnist called, "Don't-itis." A home filled with plastics would silence the harsh parental cries of "Don't!" and "Be Careful!" which magazines and parenting manuals warned could damage a child's self-esteem and cause a mother to suffer bouts of guilt and anxiety. *House Beautiful* wrote with surprising passion about the psychological benefits of indestructible table accessories, claiming that they relieved housewives "of a very real mental strain - the constant anxiety of protecting your possessions against mishap, wear and breakage. This used to ride the housewife's conscience like the old man of the sea..." *Better Homes and Gardens* told its readers that teens might even stay away from "roadside hangouts" if they could stay home without fear of damaging things. Plastic dish manufacturers took full advantage of such hopes by emphasizing the unbreakability of their products, some guaranteeing against breakage for up to ten years.

1950s teen culture was no match for this plastic armored dinette.

*Good reason why children leave home.*

Selecting child-friendly decor was considered a serious parental responsibility.

Photos of 1950s teen idols Rock Hudson and Eddie Fisher in a vinyl double frame with melamine "Mallo-Ware" tumblers molded by the Mallory Company of Chicago

The eat-in kitchen dinette was expected to withstand some rough treatment as in this depiction of mother's morning off.

Plug-in dish warmers and table side fryers were designed to create more child-centered meals with mother in constant supervision. Although also handy for entertaining, these accessories presented a constant storage headache in smaller homes.

44

Do-it-yourself polystyrene wall tiles replaced ceramic tiles for easier cleaning and quick makeovers.

Plastic dinnerware also became an important element in the postwar "life without pretense." In the new "living kitchen" of the suburban ranch home was an informal dining area handier to the dishwasher. For family dinners, a spill resistant, gay plastic dinette might be set with plastic dishes, plastic handled flatware and wipe clean plastic placemats. Using a combination of plastics and electric warming contraptions, women were expected to stay relaxed, attentive and seated during meals. These warming devices meant at least the theoretical end of getting up for seconds, while plastic dishes were to remove the threat of jagged fragments of china broken by a tearful child. Plastic dishes at these kitchen meals were also considered ideal for teaching children basic table manners. A semi-formal kitchen table could be set with plastic dishes but if children made mistakes, no expensive china was destroyed. This child-friendly feature was a strong selling point for melamine. A 1956 survey indicated seventy-three percent of melamine dinnerware buyers were between the ages of twenty-five and thirty-four and had young children.

A number of other plastic products also promised to make for happier childhoods and had the added advantage of being readily installed by a do-it-yourselfer. One such child proofing device was styrene wall tile which achieved popularity both in the kitchen and the bathroom as a kind of "suit of armor" impervious to all manner of hard use or abuse. Ads often featured cheerful mothers whishing away the crayon wall art of their children. This tile's easy maintenance must also have seemed a vast improvement over traditional walls that needed periodic painting, and it surpassed ceramic tile in its quick clean "wipeability" and ease of application. Such lightweight plastic tile was particularly well suited for the growing number of couples looking for easily installed, low cost projects that could add color and sparkle to outdated wall treatments.

Even the Fostoria Glass Company tried to prove it had products which could teach youngsters how to behave in elegant but unbreakable settings.

No other flooring material could offer as much as stain resistant vinyl in the early years of a young family's life.

DRAWINGS BY DALE MAXEY

Many homes with the popular window wall or picture window required an investment in ceiling-to-floor or wall-to-wall living room drapes.

All-vinyl floors steadily grew in popularity during the 1950s, and although their disastrous postwar entry into the flooring market slowed public acceptance, vinyl floors were gaining on rubber and linoleum floors as early as 1950. Their indisputable physical advantages were attractive to young active families where mud, food spills and rough play were part of daily existence. The most expensive floors were all-vinyl with color brilliance and abrasion resistance superior to vinyl asbestos or vinyl laminated flooring competitors.

Like plastic tiles, plastic flooring combined child proof features with a do-it-yourself appeal. Flooring "kits" were marketed to penny-conscious do-it-yourself homeowners who could create their own colorful "decorator" tile patterns secure in the knowledge that plastic vinyl would provide impact resistance, abrasion resistance, stain resistance, and easy maintenance. Even white dining area floors had become "practical" now, a symbol of the luxury that plastic promised to bring to the lives of ordinary families - even those with young children.

The difficulties presented by the sudden popularity of the window wall and picture window created another opportunity for plastic to present itself as a miracle solution for the dilemmas of new ranch house living. Modern decorating called for the open, airy feeling generated by walls of windows looking out on to a terrace. Yet to give some privacy and protection from the sun, vast expanses of expensive curtains seemed called for. How could a young family afford these expensive, perhaps custom-made curtains? And how would a busy young wife and mother find the time to do the difficult cleaning and ironing of so much drapery? Vinyl drapes were one solution to this problem, although like vinyl flooring, they suffered from a problematic beginning. But consumers were attracted to their promise of infrequent laundering and imperviousness to strong sunlight and mildew. And as printing and embossing techniques advanced, vinyl drapes lost some of their shower curtain look. They became a useful option, but only for those whose dedication to the "child proof home" bordered on the obsessive. A far more successful application of plastic in drapery was the popular white "sheer" curtain made of plastic based synthetic fibers.

Perhaps the most troubling maintenance area for families with young children was furniture. Traditional materials for floors, walls, even curtains could be washed, but what of the upholstery on furniture? Would children have to be banished to the basement or their rooms in order to protect the furniture American parents were gradually becoming able to buy? The 1950s ideal of healthy play and family togetherness discouraged such exclusive practices. The answer to this toughest of all child proofing problems came once again from the "miracle" plastic of home design, vinyl.

Vinyl drapes had the easy care properties some rueful parents needed.

No iron synthetic "sheer" fabrics performed and looked better than vinyl film in drapery applications.

Once plastic upholstery had overcome its slick, shiny beginnings, consumers looking for durability found it an excellent material for an active young family's furniture needs. One of the chief manufacturers of plastic upholstery, the Bolta Company, claimed its product was "as tender to touch as a baby's skin but takes wear like a rhino's hide." This toughness in plastic upholstery was the result of laminating a vinyl film to a woven or knit backing producing a high degree of resistance to tearing and a softer, pliable touch and appearance. When fabric backed vinyl combined with the new "no plump" foam cushions, housework was revolutionized by furniture that needed only a "quick wipe" to return to its "just like new" appearance.

In the 1950s, consumers believed they could have it all through plastics; luxury could coexist with kids and pets. The more extreme the juxtaposition, the more heroic plastics appeared as in this ad for vinyl upholstery.

47

# When their Halo slips

High-spirited children and guests were presented as threats to furniture unprotected by the satiny, no polish sheen of plastic laminates.

The toughness of vinyl upholstery also made furniture equally useful for outdoor living, a very attractive feature to the 1950s consumer. Vinyl upholstered chairs, especially those with the popular "no maintenance" black iron or wire legs, could be used on a sheltered terrace or in the living room unaffected by hard family use inside and sun, salt, cold, or rain outside.

Plastic upholstery continued to claim improvements throughout the early 1950s, and by 1956 an "elasticized" backing was added to minimize sagging or stretching. Later, vinyl upholstery was enhanced by "breathability" created by what one manufacturer described as "50,000 indiscernible microscopic pin point cells per inch" which permitted the material to "breath" like woven cloth yet keep its water repellency. These features, along with many advances in color, texture and pattern options, made plastic upholstery attractive as a living room material. Vinyl's previous confinement to kitchen chairs had ended, although it remained especially appropriate in family rooms or on dinettes where heavy wear made conventional fabrics difficult to keep looking fresh and clean. But no matter where it was used, women were told that because of it, now they could "really relax."

When plastic laminates appeared on living room furniture, a vague concept prevailed that somehow "science" had succeeded in producing almost indestructible furniture. The laminated work surface common in kitchens was now applied to the living room, particularly to the serviceable occasional tables so necessary to the 1950s concept of flexible furniture arrangements. These little cocktail, coffee or end tables were often pressed into service for family dinners in front of the TV or extra seating for guests. Their plastic "armor" could withstand cigarette burns, beverage rings, dog prints, paints, ink, crayons, scratches, and scuff marks without the need for painting, waxing or periodic refinishing. The old-fashioned methods of protecting wood using varnish or lacquer were ineffective by comparison and required periodic major refinishing. Furniture protected by a laminated plastic surface seemed to "always look like new," high praise in the 1950s.

Advertisers tried to convince consumers that child proofing one's home with plastics was a necessity for the budget and the family well-being. Yet in truth, a yearning for the beauty of traditional natural wood surfaces continued. Ads particularly stressed the ability of laminated tops to "exactly match" the real wood grain of the table underneath. The ideal child proofing product would do its work behind a mask of luxury and tradition. Coy ads for the Formica Company reflected this with the promise that their product would always remain "your own little work saving secret." This insinuation that using plastics was really a kind of secret "cheating" reveals much about the undercurrent of feelings toward both women and plastics in the 1950s. A truer understanding of the decorative potential of plastic laminates would only be realized much later by the Italian design studios of the late 1970s and 1980s.

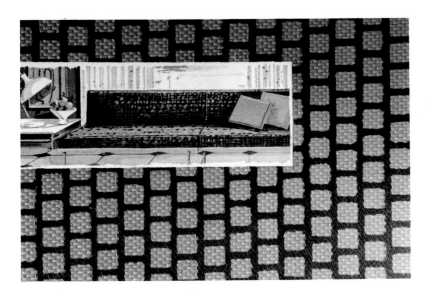

Advertisements often showed enlargements to convince consumers that "microscopic engineering" had eliminated all of vinyl's problems as an upholstery material.

"Wood grained" plastic legs completed the vertical defense of tables already horizontally protected by sheets of plastic laminates.

**49**

Informal entertaining reflected the American self-image of unpretentious friendliness.

Ads such as this one portrayed busy women as cute, perky and slightly dizzy. The serious stress in their lives was ignored except as a source of patronizing humor.

Roving guests select any seating option including the floor at this coffee table buffet party.

# Chapter 6:
## Gracious Living

With wartime rationing finally over and a naive faith in technology sweeping the country, the postwar housewife enthusiastically embraced plastics as an essential part of modern "gracious living" and its new style of entertaining. Entertaining "50s style" had been redefined as an "attitude" of irreverence, good cheer and informality. More traditional types of formal entertaining were dismissed as mere "displays" of precious and delicate things. Women's magazines advised their readers to banish all ostentatious show from their hostessing. "Nowadays you don't judge the elegance of a table by its impractical grandeur, but by its originality and appropriateness," announced *House Beautiful* in 1954.

Ironically, what was more important than fancy effects was being prepared for the "spontaneous" entertaining of unannounced visitors. Being a prepared hostess meant not only being calm and confident about all drop-in guests, but also having fun in the process. This was indeed a formidable task. But whether the constant possibility of surprise guests was real or not, readiness for it controlled many aspects of home design, making plastics an especially attractive material.

Advertisers and manufacturers of foods and tableware did their part to convince housewives that life should be a series of festive, impromptu occasions made possible by buying the newest products and foods. The conscientious housewife was told that a rather daunting number of "occasions" might suddenly confront her. Ads and articles told her to be ready to create the perfect Sunday brunch, Sunday breakfast, Sunday supper, beer buffet, cocktail buffet, cocktail hour, "night owl nibbles," "popcorn picnic," after bridge supper, after bridge dessert party, mid-morning coffee session, late evening sandwich spread, after movie snack, after theater coffee, buffet luncheon, tea party, evening get-together with the neighbors, "teen treat party," romantic twosome dinner, and "he-man barbecue." Many women might have been understandably anxious about how to execute this array of hostessing responsibilities.

Success could only be guaranteed if the questions of what to feed guests, where to feed them and on what to serve them were answered in advance. But the illusion of spontaneity was the ultimate goal at a time when women worked hard at appearing to be the gay, carefree creatures society expected. The answer to where to feed guests was the same as where to feed your family: quite simply, anywhere. In the name of "informality," women were expected to serve enticing meals from a floor to the backyard and anywhere in between. Serviceable, stackable little tables, armored in varying degrees with protective layers of plastic laminates, consequently had their heyday in the 1950s. End tables, coffee tables, TV tray tables, card tables, nesting tables, even desks and wall units that turned into tables were intended to give the hostess flexibility and her guests the freedom to roam and relax far from the "stifling" formality of the dining room.

But aside from tables, the furniture industry struggled to keep up with the new attitudes toward the role of furniture within more flexible and informal living patterns. Bound by long traditions of craftsmanship and the use of natural materials, mass production furniture makers had difficulty responding to new consumer requirements for lightweight, portable products with tough, low maintenance properties. Plastic vinyl upholstery helped update traditional forms, but did not meet the need for a new type of chair - an inexpensive, modern looking "little chair" which could be pressed into service for buffets, TV watching or barbecues, then gracefully retreat into the background.

Not providing an adequate number of little tables for your guests was considered a hostessing blunder.

**FORMICA** topped furniture is better than you need *if...*

Reg. U. S. Pat. Off.

your guests never put down a wet glass

your children never scratch the furniture

no dogs are allowed

The desirability of that "always new" look made formica a furniture essential. An unnatural sheen and slick feel are the identifiable features of this type of furniture.

A candle warmed coffee carafe with a space age star motif; Apollo Ware dishes designed by Alexander Barna

The shell chair was unobtrusive yet stylish...the perfect, serviceable "little chair." Eames side chair with "Eiffel Tower" wire base

With the new yearning for spontaneous, "lighthearted" family life and entertaining, chairs needed to be practical enough to serve a young family and mobile enough to serve the shifting groups of an informal party. For entertaining in particular, a chair had to be comfortable enough to sit in all evening, yet small enough for the smaller scaled rooms of the new ranch homes. The qualities of lightness and trimness were critical. As one writer observed, "When the party's over, what's more forlorn than a room so cluttered with chairs you have an idea the party just moved on?" These occasional chairs had to meet so many needs - for elegance, sophistication, practicality, lightweight, and diminutive size - that it is not surprising that FRP (fiberglas reinforced polyester plastic), a whole new material in a new modern form, would be the revolutionary solution.

With a trail blazed by plastics in dinettes, consumers had been exposed to plastic veneered or upholstered to furniture. But plastic had never been used structurally to build a chair until designers Charles and Ray Eames began to study a new World War II material, FRP plastic, forming it into the now famous "shell" chair.

The Eames' celebrated 1949 FRP shell chair, designed for the Herman Miller Company, embodied the mood of casual elegance and practical beauty so admired in the 1950s. With FRP plastic proving itself stronger than wood as well as weighing only a third as much, chairs in this material could stand up to large parties, rowdy teens or careless tots and still be light enough to stack and store in a closet, or move around the room to create conversation groupings, game settings or hobby areas. Besides being physically light, plastic shell chairs were visually light as well. These thin shell seats attained strength without bulk and comfort without excessive padding. And although often seen today in institutional settings like schools, offices, cafeterias, and airports, this design was inspired by the special domestic needs of the "gracious living" fifties generation.

With guests so effectively seated, the next challenge was what to feed them. For this, women were encouraged to venture into the "wonderful new world of the tinned, the quick frozen, the ready cooked." Minute rice, a can of soup and a cake mix were the short cut fundamentals of party food - buffet style. Ads and articles tried to convince skeptical housewives what rich, full lives women now had freed from the old-fashioned entertaining requirements of polishing the silver, retrieving the "good china" and slaving over a hot stove.

Those women hoping to express their individuality and creativity through cooking were urged to add their own "glamour touch" to standardized recipes found in magazines or on box labels. Packaged food was proclaimed the beginning of creativity rather than the end, with exciting and exotic meals just a matter of course since no fail mixes, like the no fail color schemes provided by magazines, were supposed to give instant confidence.

Food in a tin still carried an aura of wondrous convenience in the 1950s.

**HAM CRUSHETTES** ★

Give that leftover baked ham a brand new lease on life—with Dole Crushed Pineapple to provide the glamor! Just make your favorite ham croquettes, heat a can of Dole Crushed, and spoon it around the croquettes

Recipes based on canned foods sometimes took on bizarre forms when reaching for the "glamorous."

Candelabras, Russel Wright plasticware and a terrace setting define 1950s casual elegance.

Plastic snack trays were particularly suitable for impromptu meals in informal locations.

But even these "monograms" of individuality tacked on to standardized food were meant to be "easy" to do. What was really to occupy the housewife's mind was creating "new delights" in table setting. Not only was the housewife expected to cook a "glamour meal" but also to serve it in such a way that it would be a "memorable affair" - all the while taking care never to become part of what one writer called the "lace tablecloth school of thinking."

Memorable entertaining was to be attained by ingenuity, irreverence and surprise rather than by servants, caterers and a display of expensive china. Plastic dishes proved to be an important weapon in the hostess' unceasing fight against dull, predictable table settings and unimpressed, even if uninvited, guests. Melamine's saucy, modern appeal was far from the dreaded "rut of traditional entertaining." One melamine molder, Texasware, capitalized on this trend by gently disparaging old-fashioned china with an ad that said: "Heirlooms are wonderful but Texasware is such fun to live with." These connotations of fun and gaiety could be played up with colorful linens and a surprising dining location (a fifties favorite, the "Arabian Evening," was a meal set on a floor scattered with pillows); or plastic's gaiety could be played down with sophisticated flatware and centerpiece for the more somber "boss comes to dinner" night. Yet plasticware's essential traits of economy and durability remained unaltered.

A woman who had mastered the complexities of "spontaneous entertaining" could glamorize a can of soup with sherry or chopped chives and serve it in a sleek, trim modern plastic bowl which two hours earlier might have survived her youngest toddler. Women were promised that gay, gleaming plastic dishes could "spark the party spirit" and help put play, adventure and ultimately happiness into homemaking. The notion that a woman should be able to achieve happiness from plastic dishes and an occasional "Arabian Evening," however, proved as ridiculous and damaging as some of the exaggerated expectations set for early plastics.

But in using plastic dishes, hostesses did make it possible for buffet guests to play games or move from one conversation group to another uninhibited by the chance of stepping on or dropping an heirloom china plate. The new popularity of television, buffets and barbecuing also made plastic dishes a popular choice since meals had become portable.

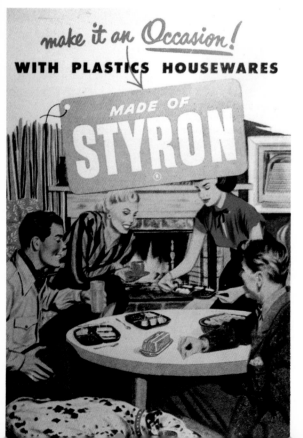

Stiff dining room rituals and heirloom china had little relevance in an era of "fun" table settings.

Light, unbreakable dishes placed on a TV tray or on a stacking stool could be carried to the family room during a favorite TV show. Or if a housewife was trying to create "drama" by serving a meal on the coffee table by the fire, on a card table by a "view window," or on a rolling cart by the patio, at least light plastic dishes carried on trays to these various locations didn't slip, slide or tilt as easily as ceramicware.

As exceptionally useful as plastics were for fifties style entertaining and TV viewing, they were even more celebrated for their place in the new "living kitchen." Gone was the antiseptic white kitchen-laboratory located in a distant corner of a rambling Victorian house, where an isolated housewife did lonely service for her family and guests. The popular 1950s ranch house style with its open floor plan created what was referred to as the "living kitchen." The clinical and sterile gave way to the festive and gay as the kitchen became a room which was visible to family and guests at all times. As a room which now had to be "decorated," the kitchen seemed an excellent place for some well-selected plastic accessories, the idea being that husband, children, pets, even guests could now keep a woman company in the kitchen thanks to the new plastic kitchen materials which were both attractive and easily cleaned.

Polystyrene snack servers by Amerline with leaf nut cups by Superlon Pro

The initial enthusiasm for the white steel kitchen with formica counters and linoleum floor faded as kitchen design evolved throughout the decade.

This scene demonstrates the concept of the "living kitchen." But although the hostess isn't lonely, her guests seem uninterested in offering much help.

**55**

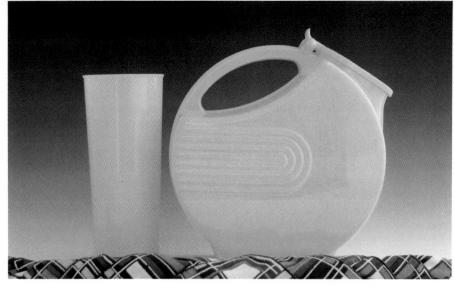

1950s plastics enjoyed unprecedented success in the kitchen - as strong as the dull and dark Bakelite of the 1930s but with a far greater color range.

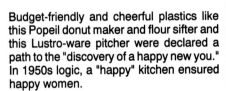

Color was the "Zing!" inexpensive plastics could provide in the kitchen.

With this goal in mind, magazines advised housewives to add a "sprinkling" of plastic kitchenware to their decor in order to "stimulate the eye" and add visual vitality without much expense. For a small investment of money, the housewife was assured that she could end her social isolation and her arduous cleaning chores as her family and friends eagerly joined her in a kitchen full of cheery, easily cleaned plastic products - walls, floor, curtains, clock, radio, counters, table tops, upholstery, knobs, handles, kitchen accessories...all of plastic.

The wide color range possible in post-war plastics was particularly important in bringing a humanizing touch to the eye-straining white enameled, chrome handled steel cabinets of the immediate postwar era. A housewife may have yearned to paint her white steel kitchen cabinets to better integrate them into the overall decorative scheme of her open floor plan. Or perhaps she lived in an apartment or an older home and dreamed of a "modern looking" kitchen. But painting an entire set of cabinets was difficult, messy and time consuming. Eventually, new formulas for do-it-yourself enamel paints made it possible to cover steel cabinets more efficiently. But plastic accessories and cabinet handles provided a simpler, more inexpensive way to change the look of the kitchen when used in combination with new curtains, shelf paper, decals, decoupage, or stencils.

Budget-friendly and cheerful plastics like this Popeil donut maker and flour sifter and this Lustro-ware pitcher were declared a path to the "discovery of a happy new you." In 1950s logic, a "happy" kitchen ensured happy women.

**56**

As the kitchen of the 1950s became an area of general living and entertaining, cheerful colors replaced antiseptic white.

When large appliance makers came out with "fashion colored" refrigerators, stoves and washers, women struggling to match these colors throughout their kitchens could count on plastic's versatile traits. Plastic accessories alone could only partially help tie together jarringly white steel cabinets with pink or turquoise appliances. But in the late 1940s and early 1950s, housewives who could not afford or could not yet attain a new kitchen in a new ranch home, looked to plastics as a way to modernize their out-of-date kitchens.

The clunky, enameled steel kitchen with its streamlined, rounded corners would eventually give way to the sleek, angular "built in" kitchen with its "sheer look" lightweight rectangular appliances of the mid-1950s. But until young families could afford these advances in kitchen design, they might settle for the vivid colors and smooth lustrous surfaces of inviting kitchens decorated with cheerful plastics.

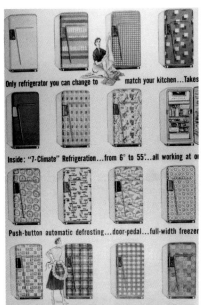

"Fashion doors" did not succeed in making refrigerators an annually purchased item, but they do illustrate that the kitchen had become a "decorated" room.

The smooth look of the "built-in" kitchen created visual unity and the appearance of increased efficiency.

57

In typical 1950s fashion, this stationary ice bucket has been given "legs" and the ability to point skyward like a rocket.

Melamine could produce curves of surprising grace, although its reputation for tough serviceability overshadowed its delicacy.

Americans with the greatest purchasing power began migrating to suburban developments in the 1950s.

Plastic was well-suited to the creation of exaggerated forms such as the popular elongated "surfboard" shape in this bread tray by Space Magic.

# Chapter 7:
# The Gay Young Modern

With a sense that the end of World War II marked the beginning of a new way of life, many Americans, striving for a more casual and spirited existence, moved into a new type of home in a suburban landscape. This break from the disasters and suffering of the recent past and an embrace of a new and different future called for a redesigning of familiar household objects. Plastics were used to satisfy this hunger for new visual forms through their vast range of design possibilities. Some of the forms created in plastic were simply not possible when tried with traditional materials like wood - or were possible only at great expense.

Plastic dinnerware designers attempted to embody this new spirit of playful modernity in forms traditional china could not match. Russel Wright's 1953 Residential line of melamine dishes surpassed the sensual organicism evident in his fabulously successful American Modern earthenware and Iroquois Casual china. For creating truly expressive flowing forms, the inherently fluid nature of plastic made it a superior material.

Some plastics manufacturers added tripod legs for that popular hovering flying saucer look, although this bowl and ashtray didn't get too far off the ground.

The Dazey Company's ice crusher was a polystyrene and metal "rocket" which ironically only worked when screwed to a wall.

The Boonton Company utilized the sense of whirling movement possible in creative plastic design by adding distinctive lipped handles to their bowls and platters.

This popular melamine bowl shape reflects a 1950s interest in oriental design.

The apparent contradiction of organic forms molded in an inorganic material did not trouble the designers of these gravy boats, one with a pony tail handle (far left).

High-sided Florenceware coupe plates; Brookpark's gently edged square plate; Boonton's squared circle bowl; an almost completely flat disc style Texasware plate

**60**

Other designers also experimented with new shapes such as the coupe plate whose flat bottom and curling edges were difficult to achieve in ceramics. Sometimes exciting formal changes came about out of necessity as in the plastic cup handle. The difficulties of fabricating the traditional looped handle found on ceramic cups resulted in Wright's much copied solution of an open hook or "pony tail" handle which flowed directly into the top of the cup.

Other design innovations were a calculated effort to target the "gay young modern," as the 1950s design-conscious consumer was known. The Boonton Company, for example, hoped to upgrade its image and customer base by introducing a daring new shape, the "squared circle" designed by Belle Kogan. This shape, and others like it, tried to express novelty and modernity without being too intimidating or austere. Adventurous forms could easily be perceived as too avante garde. In fact, Belle Kogan had agreed to "round off" her unusual square plate so that cautious consumers could continue to use their older round serving pieces while still feeling quite modern.

The smaller size of the houses built in the 1950s meant storage space could be a problem. There was no room for the old-fashioned china cabinet that held rarely used dish pieces with only single or limited functions. Plastic designers took this into account with designs which could serve double duty. A handleless sugar bowl could hold nuts, sauce, jam, or even cigarettes. An overturned vegetable dish cover could became a shallow serving platter. In such cases, "modernity" actually increased the practical appeal of the product.

The 1950s woman picked up these modern shapes in this new material to prove that the grand manner hospitality of her grandmother's etiquette books was irrelevant to her small, servantless, relaxed home. Magazine articles encouraged such innovation by reminding housewives that "...you can get the most satisfying effects if you break convention, if you surprise guests with the unusual and interesting in dishes, glassware and linens."

Sadly, selecting tableware for the creation of these "gala" dining events seemed to be one of a housewife's few means of self-expression. But plastic dishes were versatile and exciting performers in what was referred to as the "dish wardrobe" of a housewife struggling to become "famous" for her skills as an adventurous hostess.

Wifely "genius" was confined to glamourous buffets, in this case for the greater glory of a preening father and son.

"MY WIFE IS A GENIUS!"

To homemakers with butterfingers and budgets. There's many a slip 'twixt the cup and the lip—and the saucer and plate, too. But you won't have to worry about breaking dishes with this lovely, modern dinnerware.

It's molded of durable MELMAC. Brookpark is truly the economical dinnerware.

BROOKPARK
Modern design for dining

16-piece starter sets and open stock.
Beautiful decorator colors—chartreuse, burgundy, emerald green and pearl gray.

At fine stores everywhere.
Write today for descriptive folder. Dept. HB2.

international molded plastics, cleveland 9, Ohio

Despite its innovative square designs, Brookpark was initially marketed modestly as a sensible, low budget product.

Although originally it was necessary to market plastic dishes as ideal mainly for homemakers with "butterfingers and budgets," the exciting modernity represented by the forms plastic could take brought new dignity to melamine dinnerware. When Wright's Residential melamine dinnerware won the Museum of Modern Art's design award in 1953 and 1954, plasticware had achieved recognition by an elite design establishment. American Cyanamid's Melmac advertising campaign began to target bride and teen magazines hoping to make plastic dishes a longed for hope chest necessity. But the claims for plasticware's design elegance probably reached their height when *House Beautiful* awarded designer Irving Harper's Florenceware plastic buffet plate its "Classic Award" in 1955. In remarkably effusive language, the plate was described as having the "delicate beauty and warmth of the finest lacquer and the graceful proportions of a Greek vase," as well as being a good buy and a practical size for the large helpings of typical buffet eaters.

With its ability to mix adventurous designs with common sense practicality at a reasonable price, plastic dinnerware seemed to capture the essence of the American aspiration for mass produced luxury in the 1950s. Prestigious design studios like those of Russel Wright and George Nelson produced melamine dinnerware designs. Interior design magazines responded to the new material without squeamishness, as did the public. By 1960, *Good Housekeeping* magazine estimated that one out of every four American families had at least one set of plastic dishes. Decorative inlays, a vastly improved color palette and increasingly sophisticated forms made plastic dinnerware aesthetically desirable. Its practical benefits seemed undeniable.

Brookpark Arrowhead Everware, in characteristic 1950s colors of chartreuse, evergreen and gray

...it's beautiful!

.IT'S BREAK RESISTANT...IT'S THE IDEAL GIFT

Melmac is the quality dinnerware of lasting beauty.
Smartly styled in a variety of colors,
Melmac adds new excitement to breakfasts, luncheons, important dinners, yet it's guaranteed by its molders for a full year against breaking, cracking and chipping.

As plastic dishes gained consumer acceptance, images of bouncing plates and naughty children were replaced by depictions of casual but elegant settings, like this Thanksgiving dinner scene.

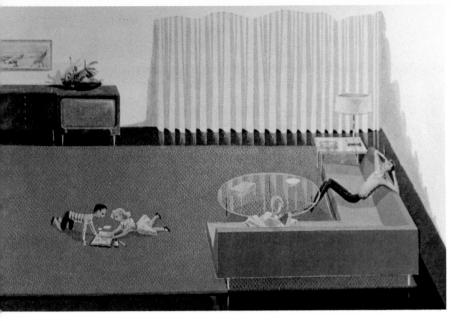

Low, boxy furniture with no-rumple foam rubber cushions floated on tiny legs to create a fresh, tailored look bordering on the austere.

In home furnishings, plastic's formal possibilities also made for exciting new design shapes. The smaller size of 1950s homes made contriving to achieve an open, airy, spacious look the overriding design mandate. This spacious look was often achieved by using wall units and sofas with long, low horizontal lines that looked serene and uncluttered under low ceilings. A trim, crisp tailored look was intended to reduce clutter and create a more fresh, clean appearance in harmony with the open floor plan and window walls.

But the trim look could become blocky and austere. The sculptural nature of plastic shell chairs with their free form multi-dimensional curves was useful in remedying this problem. Although plywood chairs molded with the help of plastic resins could achieve similar sculptural effects, FRP plastic's fluidity made possible a degree of adventurous curving wood technology could not match. (So innovative were the shell chair designs that some commentators declared them "abstract art" for the "masses" who could at last feel a bit like the art owning elite.)

A less intangible benefit of plastic's molding potential was a new capability for creating products which could be viewed from all sides. Portable chairs, radios, TVs, and phonographs, sold for their flexibility, could no longer have their backs hidden against walls. By using the sculptural potential of relatively inexpensive molded plastic, products could be used in unexpected places without creating the eyesore of flimsily encased mechanical backsides. In the case of the molded plastic chair, the back presented a contour just as pleasing and dramatic as a front or side view.

Saarinen's womb chair forms were a strong organic counterpoint to the severe geometry of trim modern furniture.

The comfortable overstuffed lounge chair continued to attract consumers even though it lacked the modern stylishness of plastic furniture.

**63**

# bubbles in a cluster...

Popular space saving bubble lights suspended from the ceiling freed end tables to do double duty as night stands, TV trays or coffee tables.

Like plastic dishes however, the formal innovations in 1950s furniture were plagued by the hesitancy of consumers to fully embrace the non-traditional forms plastic could create. Spirited modernity in design was enticing in the short term but only on a limited basis at best. The pull of traditional forms in traditional materials remained strong, and consumers were particularly reluctant to accept the thin vinyl and foam rubber padding on plastic chairs as a replacement for the comfort offered in traditional "springs and stuffing" lounge chairs.

In lighting, George Nelson's 1956 vinyl bubble lamp for the Howard Miller Clock Company was a plastic triumph, which like the Eames' shell chair, spawned many imitators. As an example of plastic's unique potential understood and stunningly utilized, the bubble lamp has few rivals. To create it, Nelson utilized a technique first employed in 1943 in a wartime effort known as "Operation Mothball" where a protective vinyl skin was sprayed over airplanes being shipped to combat duty on the open decks of ships. The vinyl could later be easily peeled off its supporting frame. But when sprayed on, it formed a weblike covering. Nelson used this technology to spray vinyl webbing over a wire "bubble" light frame.

However, most lighting design was convention-bound, and plastics were unsuccessful when used as mere substitutes for conventional lighting materials. Because despite the light transmitting abilities of many plastics, only a few heat resistant plastics could withstand the temperature of domestic incandescent bulbs. A leader in 1950s lighting design, Lightolier, did use a heat resistant phenolic plastic for the "bullet" shades of the popular space saving, flexible "pogo stick" pole light. Fiberglas reinforced plastic was sometimes similarly used. Both materials were superior to metal in their coolness to the touch and their inability to scratch or dent as painted metal might.

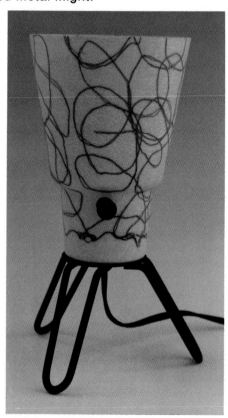

The pogo or pole light, which sometimes included plastic bullet shades, was designed to individually light each of the various activities occurring in busy open floor plan rooms.

Swirls of thread laminated over a natural "jack straw" pattern of glass fibers give this FRP TV lamp its free form look.

But pole light bullet shades were more commonly made of cheaper metals, and mass produced lamps in general excluded plastics in favor of ceramic or wire bases with paper or fabric shades. Lamps of brass, wood and linen acted as counterbalances to the synthetic look of rooms filled with naugahyde upholstery and vinyl floors. But the colors, textures and low costs possible with thermoplastics did make them popular in "mood lamps" or "TV lamps" which required only a low wattage bulb.

The injection molded polyethylene mood lamp casts a glow matching the color of its shade.

Laced or heat sealed on to wire frames, vinyl lamp shades were both convenient to clean and highly decorative.

Vinyl lamp shades did offer a limited alternative to paper and fabric throughout the 1950s starting in 1951 when decoratively patterned vinyl shades made their appearance. These shades were created by heat sealing vinyl sheets to wire shade frames or punching tiny holes into the edges of vinyl sheets which were laced to the shade's wire frame. Some shades sealed interesting decorative objects and textures like feathers, leaves, burlap, hemp, or metallic thread between two layers of plastic, creating a shade which could be easily wiped clean with a damp cloth yet still appear less synthetic because of the presence of these natural objects.

The widest use of plastics in lighting occurred outside the residential field, however. Mounted on lightweight metal frames, translucent sheets of plastic were suspended from ceilings to cover and diffuse the glare of banks of inexpensive fluorescent fixtures. This 1951 development, known as the "luminous ceiling," was immediately popular in schools, offices and businesses across the country.

Ironically, inorganic plastics made it possible for lampshades to bring a bit of the natural world into synthetic 1950s interiors. Here leaves, grasses, butterflies, and metallic flecks are sealed between layers of plastic.

The formal innovations made possible by the properties of plastics may have received a curious if hesitant reception. But there was no hesitancy in the "gay young modern's" acceptance of the color potential in home design unleashed by plastics. The enormous amount of advice offered to 1950s housewives on the subject of color indicates that it was one of the biggest anxieties of the do-it-yourself home decorator. Choosing from among the new "modern colors" was a tricky business.

"Fashion" color sold more products but created uncertainty about how to combine colors effectively.

For not only were colors more vivid (One need only look at period photos of pink living rooms and turquoise kitchens), they were also more important due to the popularity of the open floor plan. Housewives were warned that living "areas" whose colors clashed would disrupt the smooth, serene, spacious effect of a successful open plan. Colors had to blend harmoniously, yet provide "spirited accents" and subtly differentiate among the different functions of each "living area" within the open floor plan.

Even as manufacturers artificially stimulated demand by changing to ever more "up to date" and "exciting" fashion colors, they reassured housewives that no errors in color selection could occur since "pre-harmonized" schemes, charts or wheels had been "pre-selected" by "decorators" or other experts. With the proper color "key," women could move their possessions to suit the needs of the moment without fear that such mixing and matching would create embarrassing lapses in taste.

In fact, mixing and matching dinnerware was considered quite chic as well as fun to do. In 1956, *House Beautiful* wrote about this trend with a kind of amazement: "It's happening all over America - right at our dinner tables. We have been watching, for instance, the deliberate decision of hostesses not to serve the whole meal on the same kind of china. Once unthinkable heresy, today it's standard procedure."

Mixing adventurous plastic tableware designs was one way to achieve "dramatic" everyday living and "memorable" parties.

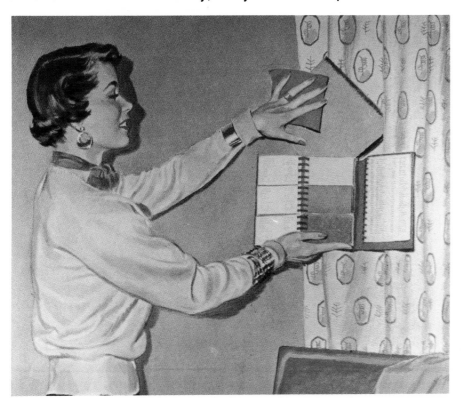

The correct color coordination for a flowing open floor plan came from studying swatches, wheels, charts, and the latest magazine advice.

Florenceware designed by Irving Harper of George Nelson Associates in 1953 for Prolon Plastics

Magazine articles instructed housewives to blend their dishes like "paints on a palette," and two-toned melamine sets allowed for this. With more of an investment, two completely different sets could be mixed. Claims were made that dishes could even be selected to complement various foods or the new living room paint job. Unquestionably, plastic dinnerware benefited as melamine's color range expanded throughout the 1950s. By 1954, George Nelson Associates' melamine line, "Florence" came in startling and dramatic "oriental colors" derived from the Japanese laquerware that inspired it. But as melamine's technology advanced, plastic dinnerware's enormous color range created new demands on the housewife to create sparkling table effects without a single misstep in taste.

Color permanence was just as important to many consumers as color brilliance. Integral color in plastic products must have seemed miraculously modern indeed. Traditional colored products that developed chips and peeling paint had to be sanded then repainted or varnished. But ads assured housewives that plastic could be counted on for its adaptability and its permanence. Since the color was "molded in," fading or wearing was said to be impossible.

Plastic's vast color range, applied to such things as radios, telephones and record players, helped make these objects into disposable "fashions" similar to the two-toned cars and refrigerators of the period. The Catalin Company, for example, told its customers it could make styrene for record players that was as "catchingly color-toned as a popular tune" and as quickly obsolete. As fashions changed, so did colors and plastic's ability to make swift, easy and cost effective color shifts made it highly effective in manipulating buyer habits.

Also appealing to a "gay young modern" was the smooth, glossy surface of many plastic products. Melamine made much of its synthetic permanency and smooth look. Although somewhat repellent by today's standards, such features were highly attractive to the 1950s consumer. The lack of fussiness in the best plastic dishes helped homemakers achieve that clean, uncluttered look so crucial in maintaining the illusion of luxurious spaciousness in a cramped suburban ranch house. Their smooth, glossy surface held out the promise that they would be like the many other plastic "miracle" products that repelled dirt "like magic." Such glossy smoothness implied easy maintenance and improved sanitation even if plastic dishes didn't deliver those things in the long run.

The easy maintenance of plastic products was in sharp contrast to this satirical depiction of Victorian housekeeping.

Picture of a slave (female variety) and her time-consuming master

Plastic dinnerware manufacturers could also boast that the smooth precision molding of each identical dish and lid solved one of the nagging technical problems in ceramic tableware. The firing inaccuracies which plagued ceramics and caused ill-fitting lids could be avoided by using the more consistent material of melamine. A similar claim was made for plastic furniture whose molded uniformity was far cheaper to manufacture than conventional jointed wood frame furniture which entailed more waste and hand work to create. This glossy modernity leant itself to affordable mass-production with very few costly finishing operations like sawing, gluing and fitting. Some shell chairs even had upholstery that could simply be snapped on and off.

The smooth curves of products molded in plastic were also particularly appealing because they had no cracks or crevices for dirt to hide in. Plastic portable record players for example, were not only easy to clean, but they also seemed a distinct improvement on other record player case materials like leather or paint that could split or peel.

But although smooth surfaces and innumerable rich colors were highly attractive, they were not the only two traits consumers desired. The yearning for a contrasting surface texture in all areas of home design became stronger as interiors became increasingly slick. The nubby texture of some fabrics, the feel of wicker strands, the gleam of polished brass - the visual appeal of these natural materials was undiminished despite plastic's advantages. Aware of this, designers attempted to simulate natural textures or give visual interest to plastics. Glass fibers in plastic shell chairs, elaborate "textures" in plastic upholstery, grasses and butterflies pressed in vinyl shades, and metal sparkles in melamine dishes are all reminders of plastic's multi-faceted capabilities despite its inherent smoothness. For to please the "gay young modern," plastic had to be a chameleon-like substance.

The machine molded lids on these melamine sugar bowls fit with an accuracy and consistency not found in ceramic products.

The focus of outdoor living had shifted to the backyard, leaving an immaculate but empty front porch.

Window walls led out to a landscaped terrace, a spot given new significance in 1950s living patterns.

Images such as this one, of an open spit barbecue in the family room, suggested to consumers that even the unlikeliest bit of the outdoors could be brought inside.

# Chapter 8:
# The Outdoors Inside

The "terrace" was assigned a particularly special position in the 1950s suburban ranch home. Meant to offset cramped interior space, it provided an extra "roofless room" with some highly romanticized functions. *House Beautiful* described this outdoor living room as "...a stage for parties, box seats for the drama of the changing seasons, an open air kitchen, a private bathing beach, and an auditorium for today's amplified systems." It was hoped that the living room's glass wall with its view of the backyard terrace would help temper the pace of modern life by bringing the restful beauty of nature within sight or use year round.

Nature's freedom and spontaneity, even if only represented by a suburban backyard, were considered powerful antidotes to modern ills. Articles appeared, some of which seem laughable today, on how to turn a carport into a patio or add a barbecue pit to the family room. New products like mobile serving carts, portable grills and electric warmers were designed to make the inside seem like the outside and visa versa. When every meal aspired to be as wholesome and relaxing as an outdoor picnic, plastic dishes seemed only natural. In this context, it is certainly true that these dishes benefited from their previous thirty year association with picnics, camping and rugged outdoor use even though to some, this rough and ready image was a dubious distinction at best.

But this craze for outdoor living meant dishes and furniture had to be modified to be able to go from house to terrace. The old "front porch style" of sagging, squeaky summer furniture and faded picnic dishes was certainly no longer acceptable. Once the terrace had been given such significance in modern life, old outdoor traditions had to be reinvigorated by new applications of strong, versatile materials such as wire, steel, redwood, aluminum, vinyl, and melamine. And now that the terrace was visible from indoors through a window wall, new respect and consideration were given to the colors and forms of patio furniture and outdoor dishes. Refinements of design combined garden airiness and freedom with indoor notions of style and taste as the boundary between indoor and outdoor living collapsed.

The unquestioning acceptance of fluid boundaries between interior and exterior living could create some odd juxtapositions as in this ad.

Bertoia's airy wire chairs were available in a coating of white plastisol with vinyl upholstered seat pads.

Retail sales of indoor-outdoor furniture soared in 1955 to 320 million dollars compared to only 90 million dollars in 1946. By 1956, black iron furniture was so popular that it began to eclipse chrome even in dinettes. Where once metal had been considered fit only for a porch glider, it now combined readily with plastics to become a furniture sensation. Black iron or wire furniture covered in vinyl upholstery could function both on the terrace or in the living room as the occasion demanded. Their plastic upholstery made possible a new standard of outdoor comfort and indoor toughness.

A less elaborate option for a purely outdoor setting was the aluminum webbed club chair, now a ubiquitous feature of outdoor living, but new in the 1950s. With its lightweight aluminum frame, this chair was easily portable. The insulating properties of its styrene or butyrate plastic arms were designed to, as one ad put it, "protect bare armed sun seekers from rude surprises." Seat and back were also of plastic in a sunfast, breathable webbing called Saran, woven out of strands of vinylidene chloride on a standard textile loom. A particularly economical feature of purchasing these chairs was the do-it-yourself kits of Saran sold to reweb frayed chairs with a still good frame.

Introduced in 1946, these aluminum and plastic webbed lawn chairs enjoyed tremendous success. For not only were they light and practical, they were easy to manufacture particularly for small scale production. Hundreds of small businesses began making aluminum lawn chairs with a garage, a bending machine, a bundle of tubing, and some plastic webbing as their only equipment.

Eames' wire chair repeated the form of his plastic shell chair, adding cut out vinyl pads for comfort.

Aluminum and plastic webbing combined to create a basic standard of comfort and light weight in inexpensive outdoor furniture design. Heavy cast iron garden furniture now seemed distinctly old-fashioned.

Plastic pools also became a popular part of 1950s outdoor living and made it possible for "everybody's kid" to enjoy what had once been a summertime backyard treat only for the rich. Plastic wading pools, along with plastic lawn ornaments and plastic webbed aluminum lawn chairs, became part of the American summer.

When Americans did venture further than the backyard terrace, advances in plastics gave picnic utensils an unprecedented degree of convenience. Increased lightness and efficiency in 1950s plastic picnic gear, particularly the styrene foam cooler of 1959, meant "roughing it" was no longer quite as rough. The styrene foam cooler was a vast improvement over previous cooler systems which sealed an insulating material between two steel shells that often dented or rusted in actual use. But ice chests made of FRP or foamed plastics were light and resisted dents and rust. And because these plastic chests had no seams or sharp corners, leaks and snagged car upholstery were avoided. Whether exploring the open road in the family station wagon or lounging at home on the backyard patio, plastics made new kinds of outdoor living possible.

THEY'RE HAPPY AT HOME WITH A
Back yard beach

If summer vacation was limited to the backyard, vinyl inflatables helped make it more fun.

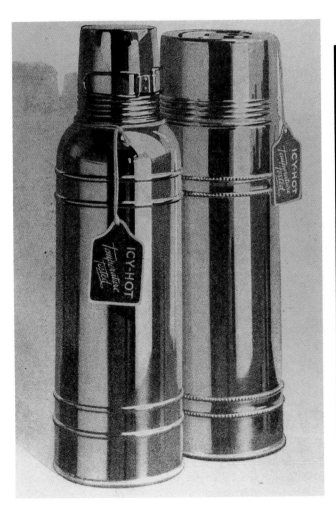

ICY HOT metal thermos from 1921

Plastic brought about the complete redesign of the thermos as thermos manufacturers gradually replaced metal parts with plastic components throughout the 1950s. In 1960, the first all plastic thermos was created.

73

After 1954, salad bowls and tumblers like those shown here, were formed using "thermowall" construction which combined plastic shells separated by an air pocket for better insulation. "Thermo" construction also meant less condensation build up and fewer water ring stains on table tops.

Plates with cup compartments tried to make buffet and barbecue eating less of a balancing act.

Metal coolers often dented, scratched and rusted - problems using plastics eventually eliminated. But plastic coolers never equaled these jaunty Scotch plaid designs in sporty stylishness.

# Chapter 9:
## Promises Broken, Promises Kept

The plastics industry of the 1950s began the decade with hopes high. Yet as the 1960s approached, much work remained to be done. Some manufacturers still did not even use the word plastic in their advertising, preferring the incomplete label "fiberglas" or the even more vague euphemism "synthetics." Plastic's reputation as an imitative substitute clung stubbornly, regardless of the stunning achievements in plastic design during the decade.

But the technical advances in creating and manufacturing so many synthetic plastics within a single generation were a clear achievement. The changes they had brought about did indeed seem miraculous and perhaps symbolic of the truly "modern" life begun after World War II. No other industry had ever expanded so fast. The period of the 1950s was one of lowering costs and constant progress in plastic manufacturing techniques and materials production. Errors and prejudices notwithstanding, plastic's astounding success was undeniable. And if consumers loudly complained about an occasional plastics failure, manufacturers could take comfort in thousands of successful plastic applications; so successful that consumers had stopped even realizing that many of the objects in their daily life - pot handles, refrigerator parts, packaging, clothespins, car interiors, hoses, eyeglasses, luggage, records, appliances, air conditioners, stockings, and telephones - were all made of plastic. Even if a product did not seem to be made of plastic, there was a very good chance that some of its most important unseen components were plastic.

Designers who designed specifically for plastic, taking advantage of its special characteristics, produced designs which remain "modern" looking even today. But although plastic fascinated the design elite, it was equally appreciated by dime store shoppers. As a truly democratic material, plastic proved its usefulness in innumerable applications for the whole range of American society.

Yet with all of its success and progress, plastic failed to be a viable alternative to many of the traditional materials it sought to challenge and replace. Plastic's supporters claimed that whenever a plastic replaced a traditional material, a betterment occurred, making plastic the new standard of product beauty and durability. Such claims, however, were less than accurate when plastic was put to the test over the course of the decade.

Polystyrene and vinyl records made plastic the standard medium for the enjoyment of music.

The "miracle" of plastic became a commonplace convenience in items such as the humble clothespin, first made in plastic in the 1950s.

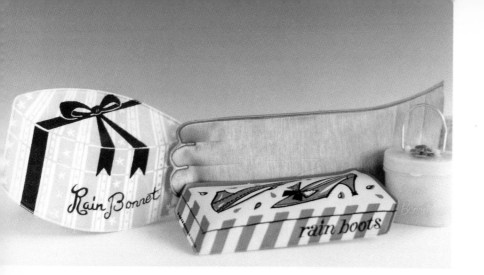

Tucked away in a purse, these clever containers held emergency plastic rain gear.

Stormette Raincoats, made of Krene, are designed with style and imagination in fashion-right colors...each comes in a handy carrying case. Women's coat about $4, some up to $8. Fashioned by Hollywood Silk Products, Inc., Los Angeles 45, Cal.

Raining beauties

MADE OF Krene®

Vinyl raincoats and booties were convenient foldaways. The reliability of their seams improved as technology advanced during the decade.

Melamine became a standard material for appliance handles like those featured on these coffee carafes from Inland Glass.

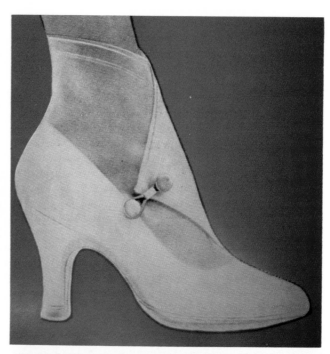

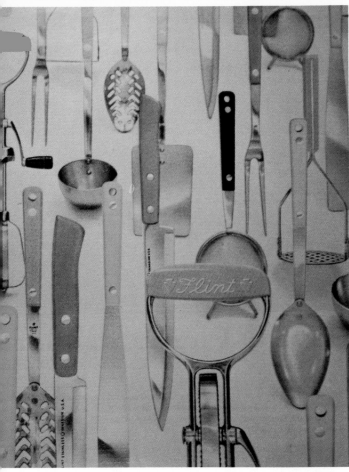

Colorful urea and melamine plastics made even utensil handles a part of the living kitchen's decorative scheme.

The plastic closet organizers we take for granted were innovations of the 1950s.

By providing more comfort, color and intriguing form, plastics made fashion history in eyeglass frames.

Despite their rattling contents and lack of privacy, plastic handbags were undeniably chic.

Vinyl teams with woven metal in this whimsical purse for theatergoers.

Heat resistant melamine and urea were the chief plastics in most quality buttons.

78

Found in tail lights as well as interior moldings, upholstery, and even experimental auto bodies, plastic became a basic material in car construction of the 1950s.

It's light! It's bright! It lives outside!

A COMPLETELY NEW TYPE OF GARDEN HOSE!

Vinyl for window shades was clearly superior to other materials in its resistance to mildew, dust and smudges.

Lighter and stronger vinyl replaced rubber in garden hoses for the first time in the 1950s.

Most of the electric kitchen clocks of the 1950s had plastic cases of injection molded polystyrene, acetate or butyrate.

The subtle side curves of this plastic kitchen clock give a touch of elegance to a utilitarian object.

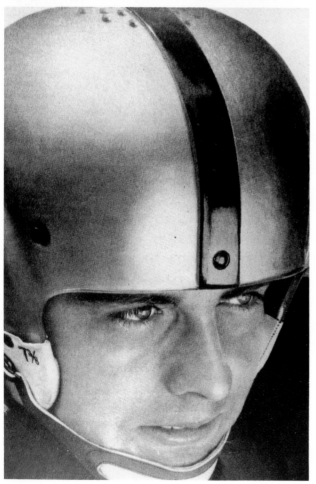

Even football helmets underwent design and safety improvements using the new postwar plastics.

Prejudice against plastic as a "flimsy" material is most obviously disproved by the hard working, long wearing plastic telephone; this style from the 1950s.

Dropping plastic dishes rarely caused a problem. Cleaning and using them did.

Despite their initial promise, plastic dishes could not replace traditional tableware. That they were nearly indestructible was true; that they were easily maintained was not. Repeated washings tended to dull and fade melamine dishes, and at least a weekly washing in a special non-chlorine sodium perborate bleach product was required to control staining. Their smooth, sleek surfaces were inevitably marred by knives even though conscientious hostesses were encouraged to plan menu items which did not need slicing on plates precariously balanced on laps and teetering TV trays.

Abandoning melamine in favor of other new plastics, however, was not feasible. In 1957, Russel Wright experimented with dishes made from the Celanese Corporation's new, more rigid polyethylene plastic called "Fortiflex" which had some of melamine's rigidity combined with an attribute melamine lacked - the capacity to store foods in the refrigerator for long periods without staining. And because it was also less rigid than melamine, polyethylene could be made into bowls with snap on lids for freezer storage. Wright dreamed of utilizing all these advantages to create a "freezer to refrigerator to table" ware whose convenience would catch on with the buying public like "oven to table" earthenware had.

With a new material and a new goal, Wright set out to expand his plastics experiments with help from a rather unlikely source, the Ideal Toy Company. With molding machines and employees idle in the toy market's annual post-Christmas slump and having already worked with Wright on creating a plastic toy rendition of his American Modern china, Ideal was selected to manufacture Wright's Fortiflex line of Idealware pitchers, leftover dishes, cold soup plates, salad bowls, and serving utensils. Unfortunately, Fortiflex polyethylene was still not rigid enough for stable dinnerware, and it lacked melamine's potential for styling elegance. Wright's Fortiflex line was a short-lived demonstration of how difficult it would be to apply any other types of plastic to the dinnerware field.

Wright's use of polyethylene in this "Idealware" design was provocative but unsuccessful.

One claim for plastic dinnerware that never really caught on with the public was that plastic material clattered less than ceramic material and created quieter kitchens. Such a benefit was of more interest to restaurants than to ordinary consumers, but eighty-five percent of plastic dinnerware was purchased for home use. This lack of interest in plastics for restaurant use proved a continuing disappointment to melamine manufacturers, particularly those hoping to expand their sales in that area once the domestic market began to sour in the 1960s. But melamine's staining, scratching and picnic associations were unacceptable to most restaurant owners, even if plastic minimized breakage expenses.

Even plastic's light weight, once such a marketable feature for casual meals, eventually became a fault as the fascination for television, buffet entertaining and outdoor living began to fade. The cult of

The creative "glamour" of franks in a skillet fizzled over time.

**Fancy Frank Fry**  *Choose your favorite fillings—*
*for a delicious change!*

informality with its craze for buffets, so appropriate for plastic dinnerware, began to lose its appeal in the endless stream of "Fancy Frank" casseroles, rickety card tables and trampled wine glasses.

Finally even plasticware's first and probably best marketed feature, its unbreakability, proved irrelevant. Although melamine dinnerware rarely broke, it did go into a gradual decay and was often discarded after a few years because of coffee stains, serrated knife scratches or dishwasher dulling.

The durability and practicality of plastic dinnerware were even more compromised by the entrance of thermoplastics into the field. As the industry became more established, dinnerware made of thermoplastics (those plastics which could be melted back into their original resin, unlike thermosetting melamine) began to challenge melamine's dominance. Melamine producers worried that such poorer quality dishes would do much to discredit all plastic dishes, even expensive melamine ones.

Such worries were well founded. The rush to get in on or hang on to plastic dinnerware profits caused lowering standards of workmanship and finish even among melamine molders. Consumer magazines reported that the plastic dishes they rated in 1964 were of much lower quality than products rated ten or thirteen years earlier. Problems ranged from unevenness in sheen and under polished edges to obvious mold lines and unsanded "flashing" (the excess plastic squeezed out when the mold compresses).

Ultimately, the absurdity and physical discomfort of the "informal" formal occasion brought about a return to the dining room table.

82

Even the quality reputation of the "Melmac" label was put into question. American Cyanamid found itself pursuing legal action against manufacturers who cut corners by using only a small amount of melamine while still claiming their dishes were "fashioned" of Melmac. Others only used melamine on certain items like cups even though labeling implied that the entire set was Melmac. And even the large china corporations like Lenox, found they had entered the plastics market too late to gather any real momentum. As the market began to fail by the mid 1960s, many china companies sold off their plastics divisions.

By contrast, kitchen plastics fared far better; for one reason the public needed less convincing. For although the 1950s idea of an attractive, welcoming "living kitchen" was warmly received, the kitchen still carried along with it the old connotations of efficiency, hygiene and service. "Industrial" materials such as aluminum, steel and phenolic plastics had historically been an accepted part of this "kitchen laboratory" concept. Because of this, postwar plastics such as polystyrene and polyethylene had far less consumer resistance to overcome than did melamine whose application to mass market dinnerware was unprecedented.

Plastic kitchenware increased both in quantity and quality throughout the 1950s. By 1956, there were at least seventy-two molders of kitchen items as the techniques for high speed injection molding improved, and manufacturing costs lessened. Consumers found in plastic kitchenware a winning combination of low cost, high efficiency and brilliant color. The extent of growth in plastic kitchenware was exemplified by one of the leading molders in the polystyrene field, Columbus Plastic Products, Inc. of Columbus, Ohio, a company which experienced

The time-honored wooden rolling pin and clay pot were two of the many objects redesigned in more easily maintained plastics.

Enthusiasm for kitchen plastics remained strong throughout the decade, although many products became increasingly utilitarian in form. Metal canisters still looked distinctly drab by comparison.

a 10,000 percent increase in its consumption of product raw materials between 1938 and 1950. In 1938, Columbus was consuming only 200 pounds of polystyrene powder a day in its production of one plastic product, a clothesline reel. By 1950, Columbus was consuming 20,000 pounds of molding powder a day to produce eighty-six items for its "Lustro-ware" line including trays, spice cabinets, cake covers, canisters, bread baskets, napkin holders, tumblers, coasters, knife racks, refrigerator dishes, and egg trays. By 1954, production jumped to over 125 articles and by 1966, the Lustro-ware line consisted of over 300 items with sales of fourteen million dollars annually.

In the kitchen at least, plastic had become an essential part of daily life by delivering the cheerful practicality it had promised. Displaced at least temporarily from the kitchenware field was metal which tended to corrode and discolor when in prolonged contact with foodstuffs. But even plastic kitchenware had to compete more vigorously with other materials when in the early 1960s, some consumers developed a nostalgia for natural materials even in the kitchen, as well as a fondness for earth colors like bronze, avocado and harvest gold. The synthetic smoothness and chemical colors of plastic kitchenware now seemed unnatural, strident, unsettling. Plastic found itself vying with calico ruffles, copper accents, pseudo "colonial" hearths, and wood grain surfaces for consumer dollars in kitchen accessory sales.

But plastics were a long way from being discredited. Vinyls in floors, lampshades, wallpapers, tiles, and upholstery did child proof homes effectively. Reinforced polyesters and saran webbing revolutionized furniture production while laminates expanded as the ideal do-it-yourself material. And portability in radios, TVs and phonographs meant impact resistant, lightweight, low cost thermoplastics were excellent product housings. Yet the inordinate fascination for these new materials in their new forms waned as the decade closed. A gradual return to formal entertaining and a renewed interest in the warmth of natural materials made plastics lose some of their appeal.

In outdoor living, more expensive indoor-outdoor furniture began to appear in natural materials. Offsetting austere black frames were sisal, wicker, rattan, and straw instead of bright vinyl. This "Pacifica look," with its evocation of tropical island breeziness, seemed more sophisticated and truly luxurious than perky, impervious plastic.

As a metal which could express strong color, copper proved a popular alternative to plastics.

In the 1960s, gold flecks and "pebbly" textures were a clear departure from the unnatural colors and smooth surfaces of 1950s flooring.

Vinyl flooring continued to be popular in the 1960s, but there remained a need to develop a more natural look in patterns and colors. The picture window and the glass wall had brought nature inside, making a muted earthy look in floors seem more appropriate. With its natural texture and color, cork flooring successfully competed with vinyl, although neither could suppress the growing popularity of wall to wall carpeting. Vinyl was more and more relegated to the kitchen and family room unless it could somehow suggest either opulence or nature.

But when plastic tried to deny its synthetic, machined identity to appeal to those who sought the natural and traditional, disaster could result. This was particularly true for plastic dinnerware whose downfall can be traced back to what, at the time, seemed like a major breakthrough. In 1953, a Swiss scientist devised the first successful foil decal for melamine dishes that was not just an easily marred surface overlay. This decal was a special melamine impregnated paper whose lithographed design would fuse into the melamine itself as it cured. It seemed that now melamine could truly compete with china even in the restaurant world, where its inability to incorporate crests and logos had hindered its advance.

By 1957, there were over 300 variations in color and pattern to choose from in melamine dishes, and indeed melamine had become much more than just a substance that could be bounced off the floor and survive unharmed. By 1964, a staggering 700 colors and patterns were available with market research estimating that seventy percent of consumers preferred decorated melamine.

But this attempt to copy the decorative traditions of china was a betrayal of plastic's synthetic, modern nature. Because plastic dishes were handled daily, they could not hide their differences from china. The traditional quality standards of dinnerware - weight, feel, delicacy, and translucency - simply did not apply to opaque, light and indestructible plastic dishes. Although some manufacturers continued to try to compete with china on its own terms, melamine sales representatives complained of being sent to the houseware department by unreceptive or disdainful buyers in the china departments of major department stores.

Eventually melamine's decorative foils made dishes capable of looking like anything but plastic.

85

Melamine could never feel like china, but simulations of china's translucency were attempted by the Boonton Company in the favorite 1950s colors of "charcoal, shrimp pink, turquoise blue and oyster white."

Look how the light shows through my new CANDESCENT

Some manufacturers, particularly china makers trying to make a go of their misguided late entry into plastics, selected shape names suggestive of a degree of luxury and status melamine simply could not embody. But despite these efforts to proclaim its stylish elegance, American buyers had never really accepted melamine for truly formal dining. Even the modern china patterns by Eva Zeisel and Raymond Loewy had never truly been accepted by average consumers, still enamored of china's traditional delicate forms.

The 1960s saw modern plastic take on "antique" pretensions as in this ad with its backdrop of misty ancient columns.

By the time this ad appeared in 1964, melamine had become largely dependent on china traditions for its design inspiration.

**86**

Dinnerware seemed to lose its focus over the decade as it moved beyond institutional food trays and toward abstract modernism, Currier and Ives scenes, even daisy flower power.

Style names like Crown Patrician, Epicure, Classic, Elegance, Debonaire, Premier, Deluxe, Regency, Hallmark, Windsor, and Imperial could not alter the facts. Melamine's light weight, chemical colors, waxy feel, and hollow sound failed to connote wealth. Surface patterns with old-fashioned, folksy names like Buttercup, Harvest Wheat, Rambling Ivy, and Shasta Daisy did not mask melamine's synthetic nature. The absurd 1964 Melmac ad campaign in which melamine dishes were displayed in china closet fashion on a seventeenth century French antique represented an ill-conceived retreat from melamine's most marketable trait, its defiant spirit of modernity.

As American buyers returned to the safety of tradition, softly colored dishes made in pretty shapes out of luxurious natural materials were the order of the day. Melamine plasticware was on its way out. No matter how many flowery, saccharine choices the endless melamine patterns offered, consumer tastes had shifted and the industry had lost sight of the source of its original success. Only a few daring explorations of the formal potential of this new material had been achieved.

Even the once revolutionary starter set selling strategy proved to have limitations. Replacement components were quite expensive and not always available, making it sometimes cheaper to start all over with a new set even if all that was needed were a few new cups or plates. And with the tremendous and constantly changing number of decorative patterns available, lines were quite often discontinued, forcing the buyer to learn a painful lesson in modern marketing.

Although some designers tried to honor plastic's potential for formal innovations, their success was limited by consumers who embraced the new and modern with short-lived zeal. Perhaps the "gay young modern" consumer would buy a boomerang coffee table and a pogo stick light. But the innovative plastic designs of the Herman Miller Company and Knoll, Inc. accounted for only a small percentage of home furniture being purchased in the 1950s. Corporations in search of a crisp, trim look were a more receptive and profitable market for plastic designs. Consequently, Knoll and Herman Miller largely abandoned the residential furniture market. What began as a plastics experiment in providing affordable, well designed home furnishings for ordinary consumers, ended as plastic anonymity in elite corporate settings.

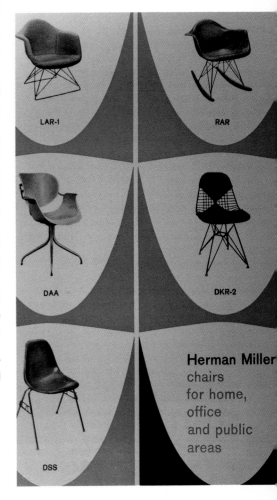

LAR-1  RAR  DAA  DKR-2  DSS

**Herman Miller** chairs for home, office and public areas

Corporate and commercial environments eventually became the home of most plastic furniture.

Cup forms that started out as innovative and rugged flared to a dramatic extreme or copied the look of the traditional china cup by the end of the decade. (left to right: Lifetime, Mallo Belle, Texasware)

At times, the formal innovation in plastics of the 1950s resulted in more and more exaggerated designs, lovely but impractical. By 1960, plastic dinnerware struggled to maintain its aura of novelty and innovation by striving for ever more delicate and thin walled forms. The melamine of the early 1950s with its thick walls and sturdy handles now seemed too institutional and too child proof. Grace and delicacy had been lost to strength. Designers and molders responded to this with thinner walls and flaring shapes. Sadly, these more dramatic forms betrayed the fundamental premise of plastic's existence in dinnerware - its unbreakability.

Ultimately, there was an inevitable swing back toward traditional materials. But after plastic's brilliant incursions, no traditional products would take their superiority for granted again. The threatening rise of plastics pushed makers using traditional materials toward improving their products, or at least their advertising. In some cases, the void left by plastic's failure spurred others to seize the market with an even more innovative material. For example, the new 1970 Corning product called Corelle "Livingware" delivered plastic's indestructibility while simulating the feel of real china. Within eighteen months, Corelleware had sold over forty million pieces.

Corning's break-resistant Corelleware filled the niche left by the declining popularity of plastic dinnerware.

Massimo Vignelli's plasticware, produced by Heller since 1969, echoes early quality melamine in its opaque colors and imaginative forms.

88

# Afterword: The Permulux Generation

As restless "gay young moderns" entered settled middle age, their priorities changed. Decorating magazines throughout the decade had offered much advice on how to create "drama" with lots of imagination and only a little money. But underlying all this decorating advice was the assumption that as soon as you could, you would "trade up" or "invest in quality" and dispense with things like contact paper, vinyl upholstery and plastic dishes. "Quality" was still defined by traditional materials of wood, glass, leather, china, and metal.

It had once been enough for Americans to aspire to wholesome fun and spirited informality. Indeed plastic had helped make these aspirations possible. But as the 1950s drew to a close, a longing for the status embodied in traditional materials and forms increased. Plastic again became a material of extremes: dismissed as a cheap dime store substitute or elevated as an avante garde material whose unique capabilities were central to the outrageous forms of the 1960s pop art movement.

Nevertheless, 1950s plastics had made a successful if short-lived break from these traditional limited roles by serving a generation's physical need for portable, adaptable and durable products. Plastic met a cultural need as well by, at least temporarily, satisfying a yearning to believe that there could be such a thing as practical elegance in the everyday lives of ordinary people. For a while at least, it seemed that plastics were bringing us progress. Light, gay and clean-lined, 1950s plastics defied tradition, making it possible for many Americans to be both "house-proud" and "home comfortable."

Kuehne 1950

Lloyd 1953

Daystrom 1953

Virginia House 1950

90

# A Collector's Guide to 1950s Plastics with Price Guide

Note: Values vary immensely according to the market location, design quality and condition of any particular piece.

## Dinettes

**1934** first chrome-plated dinette sets appear on the market
**1939** fourteen million dollars in retail sales of dinettes
**1946** fifty million dollars in retail sales of dinettes with a widening array of designs available
**1951** 225 million dollars in retail sales of dinettes; eighty-five percent of all dinettes sold use protective and decorative plastic veneer and upholstery

**LEADING DINETTE MANUFACTURERS:**
Chromcraft: St. Louis, Missouri
Daystrom: Olean, New York

As America's leading producer of metal furniture, Daystrom targeted a market they referred to as "Mrs. Home Decorator" by promising "decorator" colors, "almost daring" styling and practical upkeep in all their dinettes.

The Howell Company: St. Charles, Illinois
Kuehne Manufacturing Company: Mattoon, Illinois
Lloyd Manufacturing Company: Menominee, Michigan
Virginia House: Marion, Virginia
Virtue Brothers: Los Angeles, California

Dinettes: $300-400

Virtue Brothers 1953

*—don't worry, it's Boonton ware"...*

Dinnerware: Pieces $3-6
Sixteen Piece Set $160

| | |
|---|---|
| **1937** | melamine rediscovered by scientists studying new uses for Cyanamide at the American Cyanamid Corporation |
| **WWII** | U.S. military becomes million pound purchaser of melamine, sparking interest in melamine as a suitable material for institutional dinnerware |
| **1945** | American Cyanamid hires Russel Wright to design prototype dinnerware line for experimental testing in restaurant settings |
| **1947** | Watertown Manufacturing Company launches Lifetime Ware by Jon Hedu, featuring the coupe plate shape |
| **1951** | International Molded Plastics' "Desert Flower" by Joan Luntz, an attempt to give melamine decorative interest by incising a floral design into plates |
| **1952** | Watertown's "Woodbine" by Jon Hedu; another attempt to put pattern on melamine, this time by using a raised design instead of a recessed one |
| **1953** | Russel Wright's "Residential" line wins Museum of Modern Art "Good Design" award |
| **1953** | Irving Harper of George Nelson Associates designs the boldly colored "Florence" line produced by Prolon and inspired by oriental lacquerware |
| **1954** | Belle Kogan's "Boonton Belle" line for the Boonton Company introduces the squared circle shape |
| **1956** | Brookpark introduces the first patterned melamine, "Fantasy" |
| **1957** | Plastics Manufacturing Company molds the first "dual-colored" tableware (one color molded on top of another color in the same dish) |

## COLLECTING MELAMINE:

The physical weaknesses of melamine dishes make finding a set in perfect condition rare. High quality melamine, however, was indeed unbreakable in normal use, and because much melamine survives, the collecting outlook is a bright one.

## WORKMANSHIP:

Thick walls and rolled or rounded edges indicate melamine of the early to mid-1950's, a period when manufacturers emphasized sturdy design and high quality materials. Wary consumers were being wooed during these years, and manufacturers were taking special care to distinguish melamine as a "quality" material superior to other types of plastic used in novelty or kitchen products. Later examples of melamine may have beveled, tapered or thin edges of more grace and delicacy, but with a consequently greater potential for chipping.

## FINISH:

Finish should be smooth and highly lustrous, showing no sign of dulling, staining or scratching. The absence of mold marks, spurs or flashing indicates quality production standards and a knowledge of the manufacturing process on the part of the designer. In the best melamine, compression mold lines are skillfully hidden, and spurs have been removed by sanding.

**PATTERN:**

Some early, high quality melamine has mottled or textured cloud-like speckling, the only decoration possible before the 1955 advances in melamine decal technology. Two rare early styles used raised or recessed design to provide the decorative effect consumers expected. Otherwise, early melamine could only show its capacity to match the "decorator colors" popular with consumers of the period.

Once the decal technology for melamine was established in 1955, dishes from the late 1950s and 1960s were decorated with a huge array of traditional motifs imitative of those found on china. This trend denied melamine's nature and its modernity, factors which led to its eventual demise as a popular dinnerware material.

**CUPS:**

After 1957, some of the more expensive sets of melamine had "dual-colored" cups. Although such two tone dish styles were fashionable in general, some in the industry were actually trying to specially treat cup interiors by applying a second more stain resistant layer in response to consumer dissatisfaction which began to increase as the flaws in melamine were better understood.

Cups of the late 1950s and 60s achieved drama with their small bases and flaring tops. In use, this shape can prove unstable although aesthetically pleasing. Checking the size of the finger hole and the smoothness of a cup's edges gives an indication of the manufacturing quality and concern for function in melamine design. Handles which flow directly into the top of the cup and do not attempt to mimic the traditional china cup handle are truer to the fluid nature of plastic material.

**PLATES:**

In the second half of the decade, more expensive melamine sets had patterns on the dinner and salad plates as well as the saucer. Average or inexpensively priced sets had patterns on the dinner plates only. Designer Belle Kogan's "squared circle" style remains one of the most adventurous melamine designs along with the coupe plate. When used in an exaggerated way, however, the coupe plate's dramatic up curving sides could cause food to run toward the center of the plate. Plates of the opposite extreme, designed as flat sheets, were equally dramatic but difficult to pick up.

**SETS:**

The most commonly sold set was the 45 piece service for eight people with eight dinner plates, eight salad plates, eight cereal bowls, eight saucers, platter, vegetable dish, sugar, and creamer. Also popular was the sixteen piece starter set which was marketed and purchased most often as a June wedding gift or a December Christmas present.

1953 Woodbine by Jon Hedu for Watertown

93

## Use and Care of Melamine

For many Christmases to come — rich, well-nigh unbreakable Boontonware

1956 Boontonware

*Will it melt?*

Melamine is a thermosetting plastic, meaning it cannot be melted back to its original resinous state. It can, however, be damaged by the extreme heat of a stove. It is prudent to follow a few simple rules regarding melamine and heat sources. The ugly brown burn scars you sometimes see on melamine dishes are the result of the careless handling of melamine near a stove or barbecue.

**NEVER put melamine in the stove to preheat.**
**NEVER leave melamine next to a stove burner.**
**NEVER use melamine as a pot cover while cooking.**
**NEVER use a melamine dish as an ashtray.**

*How do I clean it?*

Melamine can be cleaned with a sodium perborate cleanser such as Dip It or a baking soda paste, for the toughest of stains. Melamine is dishwasher safe, but to protect its luster, dishes should be hand washed with a sponge or soft cloth. If coffee, tea, spaghetti sauce or berry juices are served in them, melamine dishes should be rinsed immediately after use. Ideally, all melamine dishes should be rinsed quickly to avoid surprise stains.

**NEVER use steel wool, scouring powder or chlorine bleach on melamine.(By roughening the surface, these products will only accelerate future staining.)**
**NEVER store food in melamine.**
**NEVER leave dirty melamine dishes in the sink overnight. Rinse as soon as possible after use.**

*What if it's scratched?*

Melamine scratches are usually caused by serrated knives. When using melamine, it is advisable to serve food which does not require cutting since scratches cannot be removed.

## Industry Directory

**1947-1955** Melamine of this period is generally of higher quality workmanship and finish than products made after 1955. Look for solid pastel colors, thick walls and rounded edges.
**Boontonware:** Boonton Molding Company, Boonton, New Jersey

Boonton stressed the high quality and child proof nature of its product; however, its "Boonton Belle" line, designed by Belle Kogan, did introduce the innovative squared circle plate. Boonton's involvement with melamine dates back to its molding of some of the first dishes for the military in WWII. Boonton remained a dominant and prolific producer throughout the 1950s and 1960s.

**Brookpark:** International Molded Products, Cleveland, Ohio

Brookpark "Desert Flower," designed by Joan Luntz in 1951, introduced an imbedded floral pattern into the surface of melamine dishes before the foil decal decoration was invented.

Winner of a Good Design Award from the Museum of Modern Art, "Brookpark Modern" and "Brookpark Arrowhead Everware" also designed by Luntz, make extensive use of the square shape even in cups; excellent stacking abilities; in burgundy, emerald, chartreuse and pearl gray, colors inspired by the "Far East"

The mottled look of Branchell Company's Color-Flyte

**Color-Flyte:** Branchell Company, St. Louis, Missouri

Color-Flyte was a leader in door to door sales and is an excellent example of how melamine could be given a speckled, mottled appearance to add to its decorative appeal.

**Dallasware:** See Texasware

**Devine Ware:** Devine Foods, Inc., Chicago, Illinois

As a pioneer in melamine since the 1930s, Devine Foods was the first to see melamine's advantages over metal as a food storage container material.

**Flair:** See Residential

**Florence and Prolon Ware:** Prolon Plastics, A Division of Prophylactic Brush Co., Florence, Massachusetts

Prolon was the maker of the prestigious 1953 "Florence" line designed by Irving Harper of George Nelson Associates. Harper hoped that by using Japanese laquerware as his design inspiration, his melamine line would take on a dignity which had been generally lacking in an industry which had so far only stressed the fun and practical nature of its product. His color choices of brilliant red (Sunset), flat black (Midnight), mustard yellow (High Noon), and beige-gray (Dawn) were also in sharp contrast to the usual pastels of most melamine.

Also unusual is his use of a pedestal instead of a well on which the cup rests in its saucer. As the winner of the *House Beautiful* Classic Award, the "Florence" line added to the steadily growing respect melamine dinnerware came to enjoy in the 1950s.

**Fortiflex:** See Residential

**Harmony House:** Plastic Masters, New Buffalo, Michigan

Harmony House was offered through Sears, a company which found plasticware especially appealing because its light weight reduced the mail order shipping costs of catalog orders.

**Holiday:** Kenro Company, Fredonia, Wisconsin

The name Kenro chose for its plasticware reflects the 1950s aspiration for carefree exuberance in daily living.

Prolonware cups and saucers in mix and match fashion colors

**95**

1953 Texasware

**Home Decorator:** See Residential

**Lifetimeware and Woodbine:** Watertown Manufacturing Company, Watertown, Connecticut

By introducing the coupe plate and raised decorative patterns, Watertown proved a strong innovative design force in the early years of melamine production. Its 1947 Lifetimeware, designed by John Hedu, entered the consumer market even before Russel Wright's first experimental work with American Cyanamid.

**Mallo-Ware:** P.R. Mallory Plastics, Inc., Chicago, Illinois

**Meladur:** General American Transportation Company, Chicago, Illinois

In 1949, General American molded the experimental dishes designed by Russel Wright in cooperation with the American Cyanamid Company for use in four New York City restaurant chains. Dishes made between 1949 and 1953 bear the Russel Wright signature. When Wright turned to creating his own melamine product, he sold the Meladur design to General American Transportation in 1953 with the understanding that they could continue to make the dishes but without his signature on them.

**Nichols:** Nichols Plastics and Engineering Company, Los Angeles, California

**Northern Aire Ware:** See Residential

**Prolon Ware:** See Florence

**Residential and Northern Aire Ware:** Northern Industrial Chemical Company, Boston, Massachusetts

Northern Industrial molded Russel Wright's organic melamine designs which stressed plastic's liquid, flowing nature; the way in which flash lines left by the molding process were expertly hidden indicate Wright's superior understanding of the plastics manufacturing process. Especially noteworthy is his open hook design approach to the plastic cup handle. Dinner plates have slight protrusions at each side. Wright worked with plastics throughout the 1950s and produced four dinnerware lines:

**Residential -** Translucent Lemon Ice or Sea Mist; opaque brown with dust sized copper particles (Copper Penny) or opaque black with aluminum particles (Black Velvet); Winner of the Museum of Modern Art Good Design award (1953)

**Home Decorator -** Opaque or patterned (1954)

**Idealware -** Fortiflex (polyethylene) salad, juice and beverage sets; a quickly discontinued experiment (1957)

**Flair -**Thin walled, dramatically flared shapes of oriental derivation (1959)

**Restraware and Suburban:** Applied Plastics Division: Keystone Brass Works, Erie, Pennsylvania

**Spaulding:** American Plastics Corporation, Chicago, Illinois

**Suburban:** See Restraware

96

**Texasware and Dallasware:** Plastics Manufacturing Company, Dallas, Texas

Texasware's first lines of dishes bore regional names such as San Jacinto, Rio Vista, and El Capitan; later products went under more bland names. In 1957, Plastics Manufacturing was the first to mold color on color to produce the two toned cup. The San Jacinto line, with its soft two color "clouding" effect, won a Museum of Modern Art Good Design Award (in sandalwood on white, yellow on dusty rose, white on sage green, gray on white)

**Woodbine:** See Lifetimeware

**1955-1965** The invention of a foil decal suitable for melamine created a tremendous array of patterned dishes during this time period, until by 1960, seventy percent of all plastic dishes were patterned. Quality of construction deteriorated as the emphasis on patterns increased. But a greater delicacy and grace were achieved in some design shapes through the use of thinner walls and flaring lines.

Allied Chemical Corporation, New York City, New York: *Artisan, Meladur, Sun Valley, Galaxy*

California Molded Products, Inc., Santa Paula, California: *Durawear*

C.M.P. Corporation, Old Forge, Pennsylvania: *CMP Lifetime, CMP Malibu, CMP Durawear*

Lenox Plastics, St. Louis, Missouri: *Lenoxware, Lenoxware Deluxe, Lenoxware Regency, Lenoxware Radiance, Lenoxware Americana, Lenoxware Contempra, Lenoxware Concept*

Metro Molding Corporation, Cleveland, Ohio: *Karefree, Apolloware, Golden Apolloware*

Miramar of California, Los Angeles, California: *Miramar Laguna, Miramar Imperial, Miramar Melmac, Miramar Castle*

Oneida LTD, Oneida, New York: *Oneida Deluxe, Oneida Premier, Oneida Homemaker, Oneida Custom, Oneida Design*

Royalon Inc., Chicago, Illinois: *Royalon World's Fair House, Roymac, Royalon Windsor, Royalon Hallmark, Windsor Melmac, Brookpark by Royalon*

Stetson Chemicals, Lincoln, Illinois: *Stetson Contour, Sun Valley, Stetson Riviera*

Westinghouse Electric Corporation, Bridgport, Connecticut: *Westinghouse Darien, Westinghouse Newport*

Pioneer melamine molding companies, who had set the industry standard for quality and value, became a part of the trend toward more and more patterns. Many of these patterns were floral, colonial or traditional in some way. The names selected for these melamine lines reveal a struggle between plastic's implications of modernity and a desire to express prestige and tradition.

Boontonware: *Crown Patrician, Somerset*
Brookpark: *Gaiety, Town and Country, Contemporary, Dual-Tone, Magic Carpet, Elegance, Pavilion*
Kenro: *Kenro Debonaire, Gale Art*
Prolon: *Beverly, Hostess, World of Color, Designers, Bazaar, Artiste, Potpourri, Vista, Grant Crest* (sold by W.T. Grant stores)
Texasware: *Trend, BonVivant, Marco Polo, Avant Garde, Happenings, Classics, Epicure, Park Avenue*

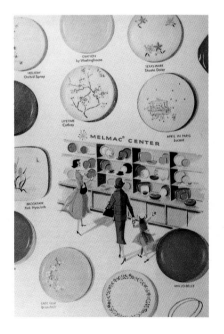

Melmac Centers displayed the wares of "Melmac" molders, all customers of melamine supplier American Cyanamid

1952 polystyrene "Chef Master" products from DaPel Plastics, Worcester, Massachusetts

These 1955 polystyrene flower pots, with drainage dish and pot that snapped together for better stability, challenged the traditional clay pot. The plastic pot was also slightly raised above the drainage dish to eliminate watering miscalculations and soggy pot bases.

Polystyrene canisters could include the kind of convenient see-through window check points metal products lacked.

*Catalin* STYRENE
Transforms Housewares into Home favorites

Remember the old fashioned bins for tea, sugar, flour, etc.? You'd usually find them hidden from view, stowed away in the pantry behind closed doors!

Today it's quite different . . . as the picture of the new LOOK* shelf-service canisters indicates. These are molded of CATALIN STYRENE in bright, cheerful kitchen colors. No bushel hides their light! Not only are they welcome, out-in-the-open show pieces but they also offer exceptional utility values and appreciated handling features. Note the transparent CATALIN STYRENE visual-inventory windows . . . a glance keeps a watchful check on the contents.

So many housewares, thru the use of CATALIN STYRENE, have been transformed into home favorites that we share the satisfaction the manufacturer enjoys from each new, ingeniously designed, skilfully molded and successfully marketed application.

CATALIN CORPORATION OF AMERICA
ONE PARK AVENUE, NEW YORK 16, NEW YORK

*Molded for Janetware Company
by The Lakone Co., 500 Rathbone Ave., Aurora, Ill.

**98**

Raffiaware salad set by Thermo-Temp in the popular "Pacifica" look with a polystyrene imitation of rope coil basketry

## *Polystyrene and Polyethylene Kitchenware*

**Burrite Ware**: Burroughs Manufacturing Company, Los Angeles, California

Plastics houseware made prior to WWII has more angles, joints and seams which are lacking in postwar plastics made by companies such as Burroughs. The smooth, one or two piece molded products of the 1950s expressed both modernity and a greater ease of cleaning. With few cracks to collect dirt and no fussy, "old-fashioned" details, kitchen plastics of the 1950s seemed especially suitable to the new notion of "gracious living."

**Gitsware:** Gits Molding Corporation, Chicago, Illinois

As makers of thermoplastic polyethylene dishes, Gits products should not be confused with the higher quality more expensive thermosetting plastic dishes made of melamine. Originally sold as "luncheon sets," these Gitsware pieces were inexpensive picnic or occasional sets not able to withstand daily use. They are not dishwasher safe and are easily marked by cutlery.

**Hemcoware:** Hemco Plastics Division, Bryant Electric Company, Bridgport, Connecticut

Offered through Woolworth's and the A&P, Hemcoware provided a budget alternative to the more expensive melamine wares; molded of urea formaldehyde with polystyrene cups.

**Lustro-Ware:** Columbus Plastics Products, Inc., Columbus, Ohio

Columbus Plastics' Lustro-Ware was the best selling polystyrene kitchenware product line until polyethylene products gained popularity in the late 1950s.

## OTHER MAKERS OF KITCHENWARE:
Beacon Products Corporation: Newton Highlands, Massachusetts
Federal Tool Company: Chicago, Illinois
Koppers Plastics: Pittsburgh, Pennsylvania

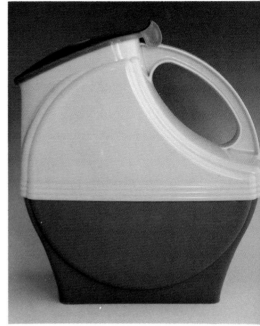

Burriteware polystyrene disc pitcher, Los Angeles, California

bread box $22
three canister set $25
ice bucket $22
pitcher $18, planter $12
salt and pepper shakers $9
ice crusher $16
clock $25

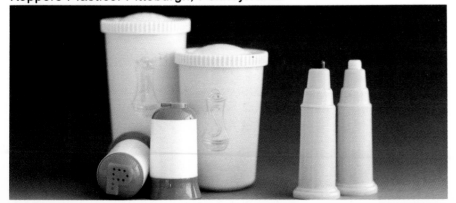

1950s salt and peppers shakers (left) in cheery, sleek shapes contrast with a pre-war set (right) in an architectural form and subdued color.

**99**

Womb sofa by Eero Saarinen

Womb Chair
with Ottoman
$900-1200

FRP shell arm chair by Charles and Ray Eames; many of the traditional labor intensive activities of furniture production, such as sawing, fitting, joining, and gluing, were eliminated in the one piece plastic shell.

Shell Arm Chair
$300-500

Shell Side Chair
$200-300

## COLLECTING PLASTIC FURNITURE:

Designs which have been in continuous production since the 1950s are of less market value than the more rare experimental or limited production pieces, which are valuable indeed. Be aware that some designs may have been reissued in recent years. Discerning 1950s originals from more recent products requires knowledge of design, color, materials, finish, hardware, and labels. Furniture designed for Knoll and Herman Miller was quickly copied by other manufacturers. Chromcraft and Selig Manufacturing were among those also producing FRP furniture in the 1950s.

### 1948  Womb or Cuddle Collection for Knoll, Inc. by Eero Saarinen

This collection was Saarinen's response to Florence Knoll's request for a chair "you could curl up in," proving furniture labeled modern didn't have be uncomfortable; womb chair, womb sofa and ottoman in fiberglas reinforced polyester resin with fabric upholstery over foam rubber.

### 1949  Shell Collection for Herman Miller Company by Charles and Ray Eames

Winner of second prize out of 3,000 entries from 31 countries in the Museum of Modern Art's "International Competition for Low Cost Furniture;" this design was first made in 1948 from stamped sheet metal with a sprayed on covering of neoprene for better warmth and a more pleasant touch. Applying FRP (fiberglas reinforced polyester) technology to the production of shell chairs interested the Eames', but mass production using this plastic was still too difficult in 1948.

FRP replaced metal in the Eames' shell chair of 1949 as a material that was lighter and warmer than metal, but just as strong. Plastic's integral color was also superior to metal with painted color which was likely to chip. The FRP plastic shell chairs were initially molded by Zenith Plastics Company of Gardena, California out of glass fiber provided by Owens-Corning and polyester resin provided by Pittsburgh Plate Glass Company. Each chair was molded in three minutes under heat and pressure. Legs were attached to the shell on rubber shock pads.

**1949  Eames Shell Arm Chair:** originally in gray beige ("greige"), elephant-hide gray, parchment, and blue-black with a solid steel rod cross base; glass fibers visible in the resin formed a natural decorative pattern, known as "jack straw;" other colors would later become available (lemon yellow, sea-foam green, red) as well as five other base options (wire "cat's cradle" base, wire "Eifel Tower" base, wooden rockers with wire base, cast aluminum pedestal base, swivel base).

**1950  Eames Shell Side Chair:** The two styles of wire cage bases were also introduced in this year.

**1951  Eames Wire Shell Chair for Herman Miller Company:**
Upholstered or partially upholstered with vinyl pad (partially upholstered chairs had triangle cutouts which exposed the wire shell to view); in tweed hopsack, black or tan glove leather, vinyl, or textured cotton; upholstery could be removed by a snap-on perimeter wire.

side chair: plastic reinforced with glass fibres. wipes clean with a cloth. use out-of-doors, too!

Eames shell side chair; portable furniture separates like these versatile plastic chairs, replaced expensive furniture suites, too formal and too expensive for the "gay young modern."

Wire Eiffel Tower, crossed rod and pedestal bases for Eames shell chairs

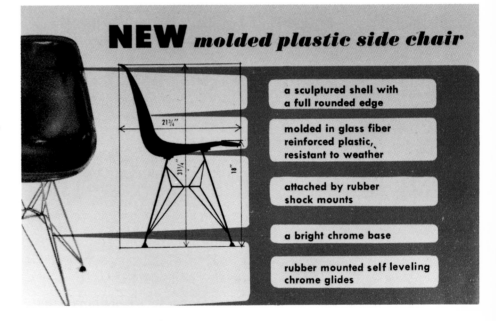

**NEW** *molded plastic side chair*

$21\frac{3}{4}''$  $31\frac{1}{4}''$  $18''$

- a sculptured shell with a full rounded edge
- molded in glass fiber reinforced plastic, resistant to weather
- attached by rubber shock mounts
- a bright chrome base
- rubber mounted self leveling chrome glides

Wire Shell Chair
$300-500

Eames wire shell chair; affordable wire shell chairs met a need for a graceful and flexible seating product. They were visually light, and air passed right through them as did rain and snow for better outside utility.

**101**

Diamond Wire Chair
$200-300

**1953 Diamond Wire Chair for Knoll, Inc. by Harry Bertoia:**
With its distinctive winglike arms and a seat cradled between triangular braces, Bertoia's wire chair is readily distinguished from Eames' shell version of 1951. Bertoia's diamond chairs came in rust proof black oxide wire or wire coated in vinyl plastisol for chip proof color and greater comfort; in two versions with the familiar low back or a high back "Bird" chair; an ottoman was also available; a detachable foam rubber pad upholstered in a rough textured cotton rayon fabric (in charcoal, camel, red, gray, tangerine, lemon, turquoise, or black) gave the buyer the option of adding more color and comfort.

**1953 Eames Shell Chairs in fabric or vinyl upholstery**

**1954 Eames Shell Chairs adapted for office use with tilt and swivel features**

**1956 Marshmallow Sofa for Herman Miller Company by George Nelson Associates:**

Produced from 1956 to 1965 with only a few hundred made, this light hearted piece is more in the spirit of the 1960s pop culture than other 1950s avante garde furniture; available in leather or naugahyde upholstery.

Marshmallow Sofa
$8,000-10,000

**1956 Pedestal or Tulip Collection for Knoll, Inc. by Eero Saarinen:**

This plastic shell on aluminum pedestal base coated in plastic was Saarinen's attempt to get rid of what he called "the slum of legs" cluttering modern interiors; because an all plastic chair was not yet technically possible, Saarinen had to use an aluminum base which he covered with a plastic coating to create the impression of a completely unified design. It was not until 1960 that a one piece plastic chair was created by Verner Panton for the Herman Miller Company.

Tulip table with
two side chairs and two
armchairs $1200-1500

**1957 Coconut Chair for Herman Miller Company by George Nelson Associates:**

A distinctive triangular chair of crisp angular appearance

Swag Leg Chair
$400-600

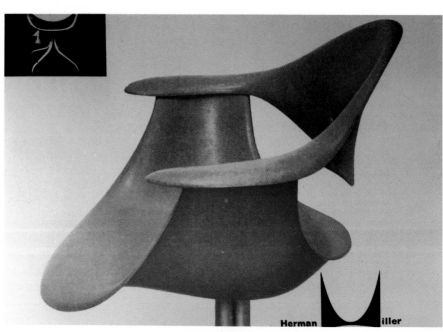

Swag leg shell chair by George Nelson

102

## 1958 Swag-legged Chair for Herman Miller Company by George Nelson:

A graceful addition to the line up of FRP shell chairs, Nelson's swag chair (with 90 degree tilting or fixed back) came in tortoise, black, pearl white, oyster gray, or combinations of those colors. It was available in dining or lounging chair heights with metal leg treatments of white, black or polished chrome.

## 1959 Eames Aluminum Group for Herman Miller Company:

(Also called the "Leisure" or "Indoor-Outdoor" group) In these designs, Eames attempted to create high quality outdoor furniture which could also serve inside if necessary; upholstery in Koroseal vinyl naugahyde or a Saran plastic weave designed by Eames and Alexander Girard; these chairs came in reclining, lounging or dining styles with optional arms and ottoman available.

Coconut Chair
$1500-2000

Aluminum Group
side chair
$200-300

Coconut chair by George Nelson Associates

### *Laminates*

Commonly referred to by the trade name Formica, laminates were discovered in 1913 as substitutes "for" "mica," the chief electrical insulating material available at that time. Laminates are formed when strips of cloth, paper or wood veneer are soaked in resinous plastic (often melamine) then dried and stacked in layers between two plate molds. Heat and pressure are applied. The resulting material can then be applied to a core substance such as wood or plywood. With both toughness and decorative potential, laminates were a popular covering for counters, table tops and furniture throughout the 1950s.

**LAMINATE PRODUCTS:**

*Micarta:* Westinghouse Electric
*Nevamar:* National Plastics Products
*Panelyte:* St. Regis Paper Company
*Texolite:* General Electric Company
*Consoweld:* Consoweld Company
*Formica:* Formica Company

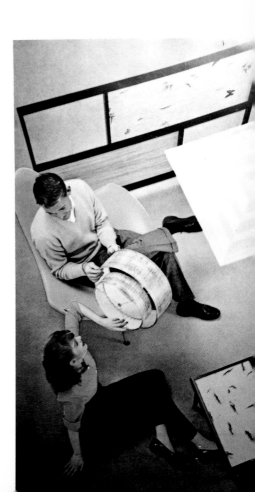

made to smile at the

*Sun*

and laugh at the

*Rain...*

Hat box $22
Cosmetic case $35
Weekender $30

## *Lawn Furniture*

Using aluminum tubing and a plastic webbing called Saran, many companies were able to provide the kind of affordable, portable and comfortable outdoor furniture still valued in similar products made for today's marketplace.

**LEADING LAWN FURNITURE MANUFACTURERS:**
The Bunting Company, Philadelphia, Pennsylvania
Deeco Inc., Huntington Park, California
Shott Furniture, Cincinnati, Ohio
Troy Sunshade Company, Troy, Ohio

## *Luggage*

1950s luggage frames were made from lightweight magnesium, plywood or fiberglas reinforced plastic. Vinyl sheets or vinyl impregnated canvas gave protection and color to exterior surfaces while rayon fabric lined the interior.

Rigid luggage, sometimes called "airplane luggage," was considered of better quality and more prestige than soft-sided bags often made of plaid material with a U-shaped zippered opening on one side. Hard-sided luggage was marketed as gender specific in its colors and shapes, while soft-sided luggage was an economical, cheery and "genderless" alternative.

Vintage luggage has predictable problems like broken hardware, broken handles, dented corners, and scuffed, stained or torn sides. Check the condition of the interior fabric, test the handle, try all latches, and examine the studs on the bottom of the suitcase which protect the bag when pulled across the ground.

**LEADING LUGGAGE MANUFACTURERS:**
Amelia Earhart Luggage: Newark, New Jersey
American Tourister, American Luggage Works: West Warwick, Rhode
    Island
Belber Trunk and Bag Company: Philadelphia, Pennsylvania
Crescent by Crescent Corporation: New York, New York
Daisy by Daisy Products Inc.: New York, New York
Durabilt Luggage, Seattle: Washington
Haliburton Inc., Los Angeles: California
Hartmann, Racine: Wisconsin
Indestructo for Montgomery Wards
J.C. Higgins for Sears, Roebuck & Co.
Koch for H. Koch and Sons: Corte Madera, California
La Parisienne, Crown Luggage: Baltimore, Maryland
Lark by Droutman Mfg. Company: Brooklyn, New York
Leed's by Leed's Travelwear Corporation: New York, New York
Mendel by Mendel-Drucker, Inc.: Cincinnati, Ohio
Olympic by Olympic Luggage: Kane: Pennsylvania
Oshkosh Trunk and Luggage Company: Oshkosh, Wisconsin
Platt Luggage: Chicago, Illinois
S. Dresner and Sons, Inc.: Chicago, Illinois
Samsonite, Shwayder Brothers: Denver, Colorado
Skymate for Samsonite
Skyway by Skyway Luggage: Seattle, Washington
Socialite by U.S. Trunk Company: Fall River, Massachusetts
Wheary Inc.: Racine, Wisconsin
Whitestar for Philadelphia Leather Goods Corporation: Philadelphia,
    Pennsylvania
Winglite by United Luggage Company: New York, New York

**104**

## *Radios*

**1949**   the first all polystyrene plastic radio cabinet
**1951**   the first clock radio
**1953**   Dow's "Styron 700," the first thermoplastic to receive Underwriter's approval for radio cabinets
**1954**   the first pocket radio, the Regency Model TR-1, weighing twelve ounces and introduced by Industrial Development Engineering Associates, Inc., Indianapolis
**1956**   the first solar powered radio

By 1950, the radio industry had substituted the new lighter, cheaper and more colorful postwar thermoplastics for cast phenolic plastic in radio housings. The miniaturization of radios, made possible by new transistor technology, also created a whole new strategy for selling radios as pocket or purse portables. Adding clocks and coffee timers to traditional table models likewise helped promote new sales in a saturated market.

**LEADING RADIO MANUFACTURERS:**
Admiral, Arlen B. Dumont Labs, Emerson Radio and Phonograph, General Electric, Magnavox, Motorola, RCA, Sylvania, Trav-Ler Radio Corporation, Tele-tone Radio Corporation, Westinghouse, Zenith

The addition of a simple handle created the "portable," a whole new way to promote flagging sales.

## *Television*

**1949**   the first plastic console cabinet, the "Consolette," thirty-six inches high and weighing thirty-five pounds, made of phenolic plastic by Molded Products Corporation for Admiral Radio Corporation; this TV was the largest plastic part that had ever been molded commercially in quantity in the United States; a three story press and 2,000 tons of pressure were required to mold it.
**1955**   General Electric introduces the first fourteen inch portable TV, a marketing and design innovation which many other manufacturers quickly copied, often using plastics.
**1959**   the Philco Predicta, an adventurous design innovation with a plastic picture tube that could either swivel on top of the console or detach completely for portability up to the distance of its twenty-five foot cord. In three models: pedestal base, table top and portable.
**1960**   "Slim" portables appear, made possible by the development of a shorter picture tube; this more shallow cabinet depth of sixteen to seventeen inches made portables look more like luggage than fat, square boxes; minimized in these slimmer TVs was the "tail" that had previously projected out the back making it impossible to put a TV flush against a wall; the slimming was heralded as an advancement in unobtrusively integrating TVs into room decor.

**LEADING TELEVISION MANUFACTURERS:**
Admiral, Dumont TV and Radio Corporation, Emerson Radio and Phonograph Corporation, General Electric, Hoffman Electronics Corporation, Magnavox, Motorola, Olympic Radio and TV, Philco, RCA Victor, Sears (Silvertone), Sylvania Electric Products, Wards (Airline), Westinghouse, Zenith

105

**MELAMINE:**

*thermosetting, colorful, tough; used in buttons, laminates, dishes.* Known as: MELMAC (American Cyanamid)

**VINYL:**

*rigid or flexible with many subtypes; used in records, hoses, flooring, auto seat fabric, upholstery, screening, drapery.*
Known as: BOLTAFLEX (Bolta Products), COL-O-VIN (Columbus Vinyl Plastic), DURAN (Maslan Duraleather Company), FABRILITE (DuPont), GEON (B.F. Goodrich), KORASEAL (B.F. Goodrich), KRENE (Bakelite), NAUGAHYDE (U.S. Rubber), RESPROID (Respro Upholstery), SARAN (Dow), TOLEX (Texileather Corporation), ULTRON (Monsanto), VELON (Firestone), VICTREX (L.E. Carpenter Company), VINYLITE (Union Carbide)

**POLYSTYRENE:**

*glasslike and rigid, with a tinny metallic ring when struck; popular in houseware and records;* Known as: BAKELITE POLYSTYRENE (Union Carbide), LUSTREX (Monsanto), PLEXENE (Rohm and Haas), STYRON (Dow)

**POLYETHYLENE:**

*tough, leathery, waxy, and very light; used in squeeze bottles, packaging, houseware;* Known as: BAKELITE POLYETHYLENE (Union Carbide), FORTIFLEX (Celanese Corporation), POLYTHENE (DuPont)

**FIBERGLAS REINFORCED POLYESTER:**

*stronger for its weight than any metal and resistant to corosion; used in pipes, luggage, coolers, boats, cars, furniture;* Known as: CORALUX (Libbey Owens Corning), LAMINAC (American Cyanamid), ZENALOY (Zenith Plastics)

**LAMINATES:**

*Strips of cotton, paper or wood veneer soaked in phenolic resins (often melamine) are squeezed in a press with heat, then applied to the top of a core substance; used for counters, tables, and other furniture surfaces;* Known as: CONSOWELD (Consoweld Company), FORMICA (Formica Company), MICARTA (Westinghouse Electric), NEVAMAR (National Plastics Products), PANELYTE (ST. Regis Paper Company), TEXOLITE (General Electric)

# Acknowledgments

Special thanks to Chris Arnold, John Berry, Ed Bickrest, Rose Cullen, Diane DiAndrea, Flying Saucer, Linda Folland, David Gerlach, Bill Graham, David Handel, Susan Hengel, William Hiscott, Anji Holtzman, Richard Kaverman, Sue Kress, Al Kress, Roy Larsen, Keith Lauer, Susan Lewin, Brenda Smith, Thomas Southall, Bob Thayer (Happy Days Collectibles), Wayne Vannatta, Joe Wood Interiors, Judith Zuidema

# Photo Credits

# Bibliography

The 1950s plastics revolution is discussed in many general periodicals of the period such as *Life, Look, Newsweek,* and *The Saturday Evening Post* as well as in "women's" magazines (*House and Garden, House Beautiful, McCalls, American Home, Ladies Home Journal, Parents, Womans Home Companion, Good Housekeeping*). Marketing and general business information is discussed in *Fortune, Business Week, Printer's Ink, Modern Packaging,* and *Advertising Age.* Trade periodicals such as *Modern Plastics, Ceramics Industry, American Ceramics, Chemical Industry,* and *Chemical Week* offer insight into the manufacturing processes of the period. Contemporary evaluations of plastic product quality can be found in *Consumer Reports, Consumer Research Bulletin, Industrial Arts,* and *Interiors.*

Bayley, Stephen. *In Good Shape: Style in Industrial Products, 1900-1960.* New York: Van Nostrand Reinhold, 1979.

------------. *20th Century Style and Design.* New York. Van Nostrand Reinhold, 1986.

Caplan, Ralph. *The Design of Herman Miller.* New York: Whitney Library of Design/Watson Guptill, 1976.

Clark, Clifford Edward Jr. *The American Family Home (1800-1960).* Chapel Hill: University of North Carolina Press, 1986.

Collins, Philip. *Radios: The Golden Age.* San Francisco: Chronicle Books, 1987.

DiNoto, Andrea. *Art Plastic: Designed for Living.* New York: Abbeville Press, 1984.

Doblin, Jay. *100 Great Product Designs.* New York: Van Nostrand Reinhold, 1970.

Drexler, Arthur. *Charles Eames: Furniture from the Design Collection.* New York: Museum of Modern Art, 1973.

Friedan, Betty. *The Feminine Mystique.* New York: Dell Publishing, 1963.

Gandy, Charles D. and Susan Zimmerman-Stidham. *Contemporary Masters.* New York: McGraw Hill, 1981.

Greenberg, Cara. *Mid Century Modern: Furniture of the 1950's.* New York: Harmony Books, 1984.

Heisinger, Kathryn B. *Design Since 1945.* New York: Rizzoli, 1983.

Hennessy, William J. *Russel Wright, American Designer.* Cambridge: MIT Press, 1983.

Hillier, Bevis. *The Style of the Century (1900-1980).* London: Herbert Press, 1983.

Hines, Thomas. *Populuxe.* New York: Knopf, 1987.

Katz, Sylvia. *Classic Plastics.* London: Thames and Hudson, 1984.

Kerr, Ann. *Russel Wright Dinnerware: Designs for the American Table.* Paducah, Kentucky: Collector Books, 1985.

-----------. *The Collector's Encyclopedia of Russel Wright Designs.* Paducah, Kentucky: Collector Books, 1990.

Larrabee, Eric and Massimo Vignelli. *Knoll Design.* New York: Harry N. Abrams, 1981.

Lewin, Susan Grant (ed.) *Formica and Design: From the Counter Top to High Art.* New York: Rizzoli, 1991.

Lisfshey, Earl. *The Housewares Story.* Chicago: National Housewares Manufacturers Association, 1973.

Matthews, Glenna. *Just a Housewife.* New York: Oxford University Press, 1987.

Neuhart, John and Marilyn Neuhart and Ray Eames. *Eames Design.* New York: Harry N. Abrams, 1989.

Powell, Polly and Lucy Peel. *50's and 60's Style.* London: Quintet Publishing, 1988.

Pulos, Arthur J. *American Design Adventure.* Cambridge: MIT Press, 1988.

Sparke, Penny (ed.) *The Plastics Age.* London: Victoria and Albert Museum, 1990.

Wallance, Don. *Shaping America's Products.* New York: Reinhold, 1956.

Wright, Gwendolyn. *Building the American Dream: A Social History of Housing in America.* Cambridge: MIT Press, 1981.

# *Index*